An Introduction to
Regional Surveying

A reduced facsimile of a mounted photographic illustration.

An Introduction to
Regional Surveying

By

C. C. FAGG, F.G.S., F.R.A.I.
and
G. E. HUTCHINGS, F.G.S.

CAMBRIDGE
AT THE UNIVERSITY PRESS
1930

CAMBRIDGE
UNIVERSITY PRESS

University Printing House, Cambridge CB2 8BS, United Kingdom

Published in the United States of America by Cambridge University Press, New York

Cambridge University Press is part of the University of Cambridge.

It furthers the University's mission by disseminating knowledge in the pursuit of education, learning and research at the highest international levels of excellence.

www.cambridge.org
Information on this title: www.cambridge.org/9781107626591

© Cambridge University Press 1930

This publication is in copyright. Subject to statutory exception and to the provisions of relevant collective licensing agreements, no reproduction of any part may take place without the written permission of Cambridge University Press.

First published 1930
First paperback edition 2013

A catalogue record for this publication is available from the British Library

ISBN 978-1-107-62659-1 Paperback

Cambridge University Press has no responsibility for the persistence or accuracy of URLs for external or third-party internet websites referred to in this publication, and does not guarantee that any content on such websites is, or will remain, accurate or appropriate.

CONTENTS

Illustrations		page vii
Preface		ix
Chapter I.	Introduction	1
II.	The Regional Survey Conspectus	8
III.	The Survey Area or "Region"	21
IV.	Maps	34
V.	The Surface Utilisation Survey	51
VI.	The Intensive Survey of Parishes	65
VII.	Transects and Relief Models	77
VIII.	The Regional Survey Atlas	93
IX.	Pictorial Illustrations (With a contribution by E. A. Robins and J. H. Pledge)	124
X.	Interpretations and Applications	133
Index		145

ILLUSTRATIONS

Reduced facsimile of mounted photograph *Frontispiece*

Fig. 1. Diagram illustrating the relationships between the various branches of regional study . *page* 10

2. Diagrammatic representation of a base map . 29

3. Specimen of 1-inch Outline Geological Survey map 40

3a. Specimen Geological Section 41

4. Specimen 6-inch base map, Trottiscliffe . . 44

5. Index to surface-utilisation categories and colour scheme 56, 57

6. Acre grid for calculating areas on 6-inch maps . 62

7. Specimen of 6-inch Ordnance Survey map numbered for field work 68

8. Enlarged plan of Downe village . . . 69

9. Specimen record cards 72

10. Specimen page of parish summary . . . 74

11. Method of projecting relief transect . . . 78

12. Method of projecting geological transect . . 81

13. Specimen vertical geological section . . . 82

14. Insertion of strata in geological transect . . 85

15. Layered relief model of Mole Gap, Surrey . 90

16. Silhouette transects, Medway Valley . . . 97

ILLUSTRATIONS

Fig. 17. Block diagram illustrating the structure of North Kent . . . . . . *page* 99

17. Specimen record of geological exposure . . 102

18. Map of Surrey hailstorm 105

19. Specimen dot-distribution diagram . . . 111

20. Map and transect, Limpsfield, Surrey . . 114

21. Transect chart 115

22. Outline of valley section 136

23. Valley section with rustic types 138

24. Valley section, Wandle Basin 141

PREFACE

A PAMPHLET entitled *An Outline Scheme for Local Surveys* was prepared by one of the authors and issued by the Regional Survey Section of the South-Eastern Union of Scientific Societies in 1920 and was shortly afterwards out of print. Since that time the authors have been collecting material for a comprehensive manual of regional surveying. They have felt, however, that while the projected manual would require several years for its completion and would necessarily be a costly volume, there is at the present time a widespread and rapidly increasing need for a less ambitious handbook which, within its scope, will give to field workers, teachers, training-college students, and senior pupils in schools some practical guidance in the organisation and carrying out of regional surveys. It is their aim in the present volume to satisfy this need.

In place of the larger work which the authors had in view it may be found more desirable to issue from time to time smaller handbooks dealing with the various branches of regional study, each being prepared by an author who combines the qualification of expertness in his subject with a full comprehension of the aims and methods of regional survey in education and research.

Lest readers who have not had experience of regional survey work should be overwhelmed by the apparent magnitude of a regional survey undertaking, it may be pointed out in the first place that in a well-organised survey many hands and many heads will contribute to the work, and in the second place that no time limit need be set for its completion. It may be found desirable at a given time to publish a selection of the material of a survey but in the nature of things the survey itself will never be completed. The great thing is that it shall be commenced and this may be done at any points which appeal to the interests of the individual workers. One of the chief aims of a regional survey is to co-ordinate all the special branches of field study in relation to a given region, and the great value of organising a survey upon a basis such as that laid down in this Introduction is that whenever a single piece of local research is accomplished it will fit into its proper place in the scheme as a whole.

Readers who are intending to engage in survey work of any kind are strongly advised to get into touch with one or other of the national societies that pay particular attention to regional survey. These are the Geographical Association, Marine Terrace, Aberystwyth, which has many local branches throughout the country; the Institute of Sociology, Leplay House, 65 Belgrave Road, Westminster; and, for Scotland, the Outlook Tower, Castle Hill, Edinburgh.

The question of colouring base maps which are printed on various qualities of paper has in the past presented some difficulty. The authors have been fortunate in obtaining the co-operation in this matter of Messrs Winsor and Newton, Ltd., who after experimenting have produced a satisfactory series of colours to their specification. These are now obtainable at a reasonable cost under the name of "Regional Survey Colours," the particulars of which are given in Chapter v.

The authors wish to express their thanks to the Council of the South-Eastern Union of Scientific Societies for permitting them to make free use of papers that have been published from time to time in *The South-Eastern Naturalist*. In the case of one such paper which is here reprinted verbatim, namely that on "Making Photographic Prints for Regional Survey", their thanks are due also to the authors, Messrs E. A. Robins and J. H. Pledge. They are especially indebted to The Director General of the Ordnance Survey, who very kindly read the first draft of Chapter IV and made many valuable comments all of which have been incorporated in the text. Figures 3, 3a, 4, 7, 8, 11, 12 and 21 are either reproductions of, or based upon, Ordnance Survey Maps by permission, and the block for Figure 19 has been kindly lent by the Royal Meteorological Society. Lastly, the authors are pleased to take this opportunity of expressing their appreciation of many helpful discussions with fellow workers in the regional survey movement during the progress of the book, and in particular with Mr Alexander Farquharson of Leplay House who has kindly rendered the additional service of reading the proofs.

C. C. FAGG
G. E. HUTCHINGS

Croydon & Rochester
1930

CHAPTER I

INTRODUCTION

A REGIONAL SURVEY may be described as the organised study of a geographical area and its inhabitants, plant, animal and human, from every aspect, and the correlation of all aspects so as to present a complete picture of the region. In such a picture we shall see the present as a mosaic, as it were, of survivals of the past and incipient phases of the future. We shall be able to interpret the life of the region as it appears to-day in the light of its past history, and in some measure to foresee and direct its future development. The scientific survey of small regions in this comprehensive sense is a comparatively modern practice, but there is a sense in which regional survey is not merely old, but primeval. We might say that the subject of regional survey is modern but that the function of regional surveying is as old as animal life; for every young organism must, according to its individual needs and limitations, investigate its environment as a part of the technique of living; and this investigation of environment, whether it be by caterpillar or child, is the very essence of regional survey. It is also the beginning if not the end of education.

We shall see in Chapter III that the essential feature of a region, in the sense in which we use the word, is a social nucleus or centre of population. To define the limits of a region is often a difficult matter, but there can be no doubt about its nucleus, whether it be village, town, city or metropolis. The smallest social nucleus is a home, just as the smallest social unit is a family, and for an exploring infant its home constitutes a complete region. It is well to bear in mind that even the most elaborate regional survey has its prototype in the child's investigation of its immediate surroundings. The regional survey movement of the present century is in reality a highly developed manifestation of this same primitive or infantile function of exploration. We are thus all regional surveyors whether or not we are conscious of the fact, and our effectiveness in the world depends in large measure upon our efficiency in this rôle. We may well illustrate this view by considering the case of a family removing to a new residential district. Let us suppose that Mr Jones of Liverpool, for business reasons, has to find a new home within reach of the metropolis. He has been advised by a friend that Oxham is a desirable neighbourhood and he and his wife travel thence to make a reconnaissance of the place and its amenities. They are charmed by the general atmosphere of the locality, Mrs Jones is favourably impressed by the shops, they discover that the town contains suitable schools for the children and that the train service to London is convenient. The next step is to find a suitable residence and with this quest a more intensive survey of the neighbourhood commences. Being wise people, they will take advantage of some of the survey work that has been performed by others, particularly that which has been published in the form of a map of the neighbourhood, and that which is represented by the knowledge to be obtained from local estate agents.

Having found a house they eventually move in and subsequently each member of the family embarks upon a more leisurely survey, extending the limits of the region as time goes on. It would be intriguing to follow out in imagination the images of their new region which they severally build up, and to place the elements of these images into the scientific categories with which the regional surveyor is concerned. We can, however, only pause to note that each would form a different image limited by his or her interests, observations and experiences. The region is infinitely more than any of these images of it, or all of them together; more than all the images that have ever been formed of it, more indeed than the completest human survey can ever discover and describe. But the recognition of our limitations need not daunt us in our efforts. The insight that unexpectedly but regularly comes to those who pursue regional studies *in the field* is, in itself, an ample reward for their labours, apart from the value of their work for others.

A regional survey may be undertaken from any of three motives or all of them. It may be regarded as a piece of pure research, as a method in education, or as a preliminary to a scheme of civic improvement or development, e.g. a town or regional plan.

This introduction is addressed particularly to those who are interested in regional research for its own sake and to teachers who wish to use it as an instrument in education. What we may call the amateur survey and the school survey are essentially the same, for regional research is an extended education for those who participate in it, while a survey of the school surroundings will bring the spirit of exploration into the work of the children, and often produce results that are worthy of the title of research. The products of the two types of survey will, of course, be judged by somewhat different standards, and in a school survey almost the same ground will be repeatedly traversed by successive groups of pupils; for it cannot be too strongly emphasised that the educational value of regional survey work lies in the *making* of surveys and not in the study of those which have been made by others. Again, whereas the workers in a co-operative scheme of regional research will make the fullest and freest use of all recorded information about the region, such information will only be disclosed to pupil surveyors at the discretion of the teacher. For example, while elementary pupils may fittingly be required to make scale plans of the school premises and their immediate surroundings by actual measurement, an advanced surveyor would save himself this trouble by procuring the best available maps. He will make his own plans, e.g. of earthworks, only when satisfactory plans are not already obtainable. All this is not to say that a school survey will not be a progressive undertaking; on the contrary, each fresh group of pupils, if they are allowed sufficient scope, will discover new aspects of their region, and the honour of adding something to the school's permanent collection of material will prove an inspiration to the children. Still more will this be the case if a research survey of an area including that of the school is in progress. In this case the senior pupils should be encouraged to produce work worthy of addition to this survey, which will often be housed at the library for public reference. We may add that we have seen several school surveys of which more mature survey organisations might well be proud. It is necessary, however, to guard against over-emphasising this aspect of school surveys. Their educational value will be seriously impaired if encouragement is not also given to those pupils whose work is not up to exhibition standard. The teacher will, as a matter of course, endeavour to get from every pupil the best work of which he or she is capable, but in regional survey the cruder efforts of the less competent children are of even greater benefit to them than their more encouraging results are to the brighter pupils.

It will not, we hope, be deemed an impertinence if we urge upon advanced workers the importance of paying heed to the mode of presentation of their results. Those readers who already realise this, will appreciate the need of so doing. Much, if not most, of the world's best work in scientific research has been accomplished by amateurs, but *amateurishness* in the presentation of results should as far as possible be avoided. Good field work is often spoilt by being presented in slipshod fashion on maps, or by long and wordy descriptions instead of concise or even tabular or graphic statements. As a hobby, regional survey merits the expenditure of at least as much pains and money as, say, golf, photography or music. The regional surveyor should emulate the devotees of these cults by obtaining the necessary equipment and acquiring some skill in using it. If, for instance, one is going to spend some hours in preparing a map to show the results of some months of study it is worth spending a few pence upon a sheet of hand-made paper for the purpose, instead of using paper that will acquire a dingy appearance in the course of a year or two. It will of course often happen that the time of a first class field worker can be more profitably spent than in the more or less mechanical work of draughtsmanship, or he may have little aptitude for it. For this reason, and for the often desirable work of duplication, the services of one or more capable draughtsmen are invaluable to a survey organisation. A survey which pays heed to these matters, though its intrinsic merits may not be greater, will command infinitely more respect than one which does not.

We may now pass on to the brief consideration of the survey as a preliminary to civic improvement or development. The scope of such a survey will necessarily be restricted more or less to the requirements of the particular purpose in view; otherwise the realisation of that purpose will run the risk of being indefinitely postponed. It is, however, an axiom of town planning that a plan which is not preceded by an adequate survey is worthless. "The object of Town Planning", writes the Chief Town Planning Adviser to the Minister of Health, "may be briefly described as the direction of all development so that all land may be put to the use for which it is best fitted, the health, wealth and happiness of the community being paramount considerations....Town Planning is always a balance of considerations, and therefore it is obvious that all the facts of the situation, their causes and relations, should be before the planner."[1] In the urgency of preparing a plan, however, the town planner is rarely able to make so comprehensive a survey as he would wish. He will count himself fortunate if the town for which he is called upon to make a plan is the centre of a regional survey organisation; for not only will such an organisation have gathered together a bibliography of all sources of information regarding the town and its surroundings, but it will have pursued its own inquiries in at least some of the directions that are of special value to him. Further, although the survey activities of a group of research students, or of school children, should not be "limited to items that have a clearly demonstrable practical value", the needs of constructive planning may temporarily direct those activities into such channels.

[1] G. L. Pepler, "Regional Survey as a Preliminary to Town Planning", *South-Eastern Naturalist*, 1925.

We may fitly close this chapter by expressing the ideals of a regional survey in the words of a far-seeing regionalist. "Let us study life regionally. Let us see the town as the focus of spiritual expression. Let us try to plan, not merely to meet problems of overcrowding, but in such a way as to provide the means of maintaining all that is of vital value in the inherited tradition of the town. Let us help present tendencies towards that closer co-operation and communal action which may once more work a spiritual regeneration of our towns and of our civilisation."[1]

[1] Professor H. J. Fleure, "The Regional Survey Preparatory to Town Planning", *Town Planning Inst.* vol. IV, No. 3, 1918.

CHAPTER II

THE REGIONAL SURVEY CONSPECTUS

"All the sciences are connected; they lend each other material aid as parts of one great whole, each doing its own work, not for itself alone, but for the other parts; none can attain its proper result separately, since all are parts of one and the same wisdom."

ROGER BACON, 1266.

THE objects of a conspectus are threefold. In the first place it should serve to show at a glance the whole scope of a regional survey and the relationships which exist between its various branches. Secondly, it should provide a sufficiently detailed summary of the whole field to enable workers in a co-operative undertaking to pick out the subjects in which they are particularly interested, at the same time showing the places occupied by these subjects in the general scheme of regional research. Lastly, it should supply a classification which will enable a society or school to arrange all its material, whether maps, graphs, tables, written matter or pictorial illustrations, in a suitable order and provide an index to them.

The diagram (Fig. 1) is an attempt to fulfil the first of these functions. If the happy child's joyous curiosity concerning its environment survives the ordeal of examinatory education, he will of necessity pursue his studies in one or more of the branches into which field research has been divided. He will, however, be severely handicapped and the value of his work seriously limited if, as too often happens, he specialises to the extent of ignoring any or all of the aspects of nature which appear to be outside the scope of his particular "ology". The diagram shows the various branches of field study arranged in an orderly sequence, thus providing the logical basis for the extended conspectus set out below. It also indicates concisely the ways in which the groups of phenomena represented by the various sciences act and react upon one another to produce the drama of the region. It has been found very helpful in teaching the interdependence of the field sciences, particularly when it has been gradually built up on the blackboard in successive lessons in conjunction with a transect diagram of the locality constructed as described in Chapter VIII.

At first sight the diagram may appear rather formidable, but upon closer examination its essential simplicity will emerge. For teaching purposes it may be either simplified or elaborated according to the needs of the pupils. It should be read from below upwards, following the main headings in numerical order, and observing the arrows and the notes upon them which indicate the direction and manner of the influence of each group of phenomena upon the others.

The groups one, two and three represent the physical features of the region. Of these the earth's crust is literally its foundation and the study of the earth's crust, which constitutes the science of geology, is of primary importance in a regional survey. Different parts of even the smallest region will usually be found to exhibit different kinds of rock.[1] One part may be on clay, another on sandstone or gravel and yet another on chalk or limestone, while in the northern and western parts of

[1] The colloquial use of the term "rock" is somewhat narrower than its use in geology where it means mineral matter of any kind occurring naturally in large quantities. Thus, to the geologist a soft clay or a loose sandy formation is just as much rock as limestone or granite.

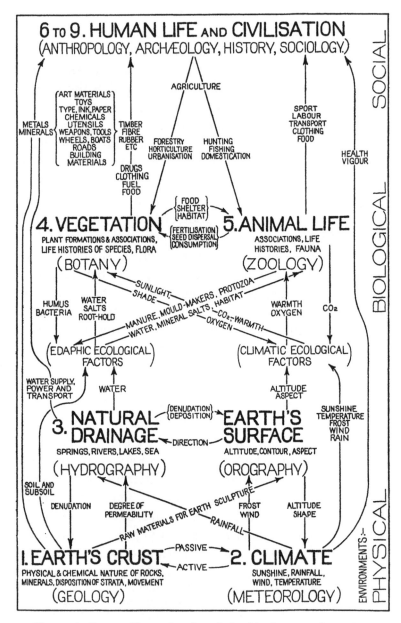

Fig. 1. A diagram illustrating the relationships between the various branches of regional study.

Great Britain the igneous rocks will often be represented. It will be necessary to observe what rocks occur at the surface, and to examine and record their physical and chemical nature. Their physical characters, that is, whether they are hard or soft, compact or loose, impervious to water or permeable and so forth, are physiographically of greater importance than their chemical nature. From the economic standpoint, however, their chemical constitution and particularly their mineral contents are often of the greatest importance. With certain exceptions in the case of igneous rock, the formations will be found to be disposed in layers or strata lying one over the other as seen in the section on p. 81. The strata are usually more or less tilted, often displaced by what are called faults and at times much contorted. The way in which the strata are disposed to one another is, as we shall see, the key to many of the physical features of the region. The fossil remains of organic life, if any, to be found in the rocks have a story of their own to relate and they should not be neglected in a regional survey. But, with certain exceptions, the fossils as such have no great significance in relation to other branches of a regional survey.

We shall return to a fuller discussion of geology and of each of the other aspects of the region in Chapter VIII; for the present our aim is to obtain a comprehensive view of the whole field. We thus see in our region a portion of the earth's crust, composed of various rocks and with definite structure, standing above the level of the sea. This forms the raw material for earth sculpture, offering a passive resistance to the attacking forces of our next group of natural phenomena, known collectively as climate, the scientific study of which is called meteorology. Not entirely passive, however, for probably no part of the earth's crust is absolutely stable and if the region includes a portion of the sea coast the rising or sinking of the land, however gradual, may be of considerable importance for our regional studies. But apart from this and the occasional occurrence of earthquakes we may regard the earth's crust as a passive factor.

The main elements that go to make up the climate of the region are sunshine and temperature, atmospheric pressure and wind, clouds and rainfall. These, of course, are all dependent upon one another and together they make a steady attack upon the earth's crust tending to reduce it to the level of the sea. The sea itself, aided by tidal action, participates more conspicuously though not more potently in this attack, and the study of coast erosion will form an important part of the survey in a maritime region. Of the agents of climatic or sub-aerial denudation the rainfall is by far the most important. Its interaction with the earth's crust gives rise to our third group of phenomena, namely the natural drainage of the region, that is its springs, rivers and lakes. If we look in the diagram for all the arrows directed towards "Natural Drainage" we shall see all the factors which help towards its determination. From the climatic group the rainfall provides the raw material which is converted, as it were, into drainage systems by the earth's crust. The rain which falls upon the earth's surface meets with a different reception from each kind of rock. Some of it will remain in the soil to be re-evaporated either directly or through the medium of vegetation. Of the remainder, that which falls upon impermeable rocks such as clays will either rest upon the surface until re-evaporated or run off in directions determined by the slopes. A great part of that which falls upon porous or jointed and fissured rocks such as sands, gravels, sandstones or limestones, will percolate until it reaches either an underground impervious stratum or the level of saturation, to re-issue in the form of springs at favourable points. In either case it will eventually find its way by rills, streams and rivers to the ocean whence it came. On its way it will effect a certain amount of denudation of the earth's crust, removing lumps or particles, in suspension or in solution, from higher to lower positions, or to the sea, for re-deposition. These processes give rise in time to modifications of the earth's surface which in turn may react upon the climate itself and change the incidence of the rainfall. The arrow pointing from natural drainage towards human life should in truth point backwards also; for man, the greatest of all regional exploiters and developers, with his canals, reservoirs, conduits, dams, sewers and wells, often exerts a profoundly modifying influence upon the natural drainage. Even the vegetation and animal life, though in infinitely less degree, are not without their effects in this direction.

We may pause here to note that the exact sciences of mathematics, physics and chemistry, though they do not figure in the diagram, deal with natural laws which are operative throughout. These sciences are the abstract forms of the concrete studies mentioned above.

All the phenomena so far considered constitute the physical features of the region. Together they furnish the factors which determine the types of vegetation, so far as these are physically determined. These factors have been divided by ecologists into two groups, namely, the edaphic or soil factors and the climatic factors. In dealing with vast continental regions and in mountainous districts the climatic factors are of major importance. We speak for instance of tropical and temperate vegetation or of the arctic-alpine flora. But in the small "regions" with which we are concerned, all of which fall within the same climatic zone, the edaphic factors become the basis of our classification of types of vegetation, while the local variations due to climatic differences are usually of minor importance. Looking again at the arrows we may note that the vegetation obtains water, salts and roothold from the soil. The climatic factors include sunlight, warmth and carbon dioxide, to which may be added humidity and wind, the latter with its stultifying effect on the one hand and as an ally in seed fertilisation and dispersal on the other. The animal life is very intimately associated with the vegetation and affects it by consumption, and by assistance with manure and in the fertilisation and dispersal of seeds. Lastly, mankind with agriculture, forestry, horticulture and urbanisation exerts a dominating influence upon the plant life of the region.

The animal life is in the first place dependent upon the vegetation for its food and for shelter and habitat. It is also conditioned in some measure by the geological and climatic factors and, as in the case of plant life, the competitive and co-operative activities of the animals amongst themselves are of much importance. Here again mankind plays the dominant rôle in civilised regions. By agriculture and fishery, breeding and domestication, preservation and extermination he changes the constituents of the fauna more or less at will.

Lastly, at the top we come to the consideration of human life and culture, the understanding of which is indeed the final object of a regional survey. Not that we should consciously limit our investigations in other branches to those aspects which appear to have a direct influence upon human life, still less to those inquiries which have a directly utilitarian value. The content of the human studies, including sociology, economics and history as well as anthropology and archaeology is so extensive that we have given it the remaining four numbers (6–9) to correspond with the notation of the extended conspectus. The reader will be able to see by again observing the arrows some of the ways in which mankind is dependent upon his physical and biological environment. Enough has already been said for the present of his active exploitation and development of these.

The detailed conspectus set out below and the notation applied to it are intended to meet the other two needs mentioned at the opening of this chapter. The discussion of the separate items in it we shall postpone until we reach Chapter VIII. The notation is an adaptation of the decimal system devised by Dr Dewey in America and used by most public libraries in this country. Before describing the system it will be well to meet the objections that are sometimes raised regarding it. In the first place it should be observed that there are two distinct parts to the Dewey system, namely, a classification of all the subjects that anyone has ever written about and the application to this classification of a decimal system of indexing.

Whether or not Dewey's classification is the most suitable for dealing with the general chaos of literature need not concern us here; it is totally unsuited to the purpose of dealing with regional survey material. We have accordingly prepared a special classification, based upon the foregoing diagram, which takes account only of categories relevant to our purpose. The classification here given was first published in a paper entitled "An Outline

Scheme for Local Surveys" in the *South-Eastern Naturalist* in 1920. Perfection is not claimed for it but it was adopted only after draft copies had been circulated amongst leading regionalists and specialists in the various fields and after a committee had given very careful consideration to their replies and as far as possible incorporated their suggestions. Since its adoption it has been in constant use by the authors and others for arranging and indexing maps, written matter, photographs, lantern slides and other survey material and found, with the slight modifications now incorporated, conveniently to meet all requirements. Individual surveyors or survey organisations may wish to make minor alterations to meet special needs. On the other hand, the advantages of the adoption by all surveys of the same classification and arrangement of material so far outweigh the defects inherent in any such classification, that a fairly close adherence to the one here given is strongly advocated.

The connection of our scheme with that of Dr Dewey is thus reduced to the application of the decimal indexing notation. The only criticism of this by those who admit at all the need for a notation is that the subdivisions of knowledge do not naturally and invariably fit a tenfold classification. It should be observed, however, that the only limitation of the system in this respect is that any given category cannot be divided into *more than* nine subdivisions of the next lower order. This limitation is chiefly theoretical for more than nine subdivisions are very rarely required. Moreover, it happens that the main branches of regional study can be very conveniently arranged in nine categories, a tenth being thus available for general results and questions of method. In the secondary and subsequent subdivisions there is, of course, no need to use the whole of the nine available numbers if they are not required and it will be seen that we have appropriated only seventy-five out of a possible ninety numbers.

The system of numbering is as follows. The figure on the extreme left indicates to which of the ten main divisions the document bearing it belongs. Thus the "4" in 430 indicates the main division *Vegetation*. The figure next to the right signifies a subdivision of one of the ten primary divisions. The "4" in 540, for instance, denotes the subdivision *Subterranean Animals* of *Animal Life* (500) or in 740 the *Late Mediaeval* period of *History* (700). The third figure from the left denotes a further subdivision of the subject indicated by the first and second and so on for the fourth, fifth, etc., as far as we choose to carry the process of analysis. Except in the case of *Methodology, etc.* (000) the significance of the third figure has been left open in the conspectus. Each survey society can make what use it chooses of the third and subsequent figures without interfering with the general scheme of arrangement. Subdivisions of the third order may well be varied in different districts. For instance, under *Stratigraphy* (110) in the south-eastern counties we use the nine digits in the third place to denote the various geological formations that occur there, but a survey being conducted in, say, Wales would have an almost entirely different set of geological formations to deal with.

In the case of *Methodology*, the third figure will have the same significance as the corresponding first figure in the nine main divisions. The cipher always denotes the "general" aspects of the subject represented by the figure to the left of it. Thus, the number 030 on a document would show that it related to maps in general. It would be applied, for instance, to a catalogue of Ordnance

Survey maps or to Chapter IV of this book. A document marked 031 would contain information concerning geological maps while a paper on, say, the interpretation of symbols on Elizabethan maps would be marked 037.6. In some cases subdivisions may be shown by cross references. Thus, 710/870 would indicate Roman roads. Documents which, like this volume, would be indexed under numbers between 020 and 099 are part of the surveyor's equipment rather than part of the survey. They will not of course find a place with the material of the survey.

THE EXTENDED CONSPECTUS AND NOTATION

000 General Results and Interpretations of the Survey.
  010 Geographical Position and Environs of the Region.

METHODOLOGY AND TECHNIQUE OF SURVEYING.

  020 Books, Maps, Records, Illustrations (Bibliography).
  030 Information concerning Maps.
  040 Methods of Surveying Rural Areas.
  050 Methods of Surveying Urban Areas.
  060 Methods of Surveying Maritime Areas.
  070 Methods of Recording, Illustrating and Exhibiting Results of Survey.
  080 (Open.)
  090 Theories and Methods of Interpretation.

100 The Earth's Crust (Geology).
  110 Strata of the Region and their Sequence (Stratigraphy).
  120 Geological Structure of the Region (Tectonics).
  130 Physical and Chemical characters of Rocks (Petrology).
  140 Mineral Contents (Mineralogy).
  150 Fossil Contents (Palaeontology).
  160 Economic Geology.
  170 Soils and Subsoils.

200 Atmospheric Phenomena (Meteorology).
  210 Pressure and Wind.
  220 Rainfall and Humidity.
  230 Sunshine.
  240 Temperature.

THE REGIONAL SURVEY CONSPECTUS

300 Surface Features and Natural Drainage (Orography and Hydrography).
- 310 Relief and Bathymetry.
- 320 Rivers and River Basins.
- 330 Lakes, Ponds, Bogs, Marshes.
- 340 Underground Drainage and Springs.
- 350 Erosion and Deposition.
- 360 Sea Coasts.

400 Vegetation (Botany and Plant Ecology).
- 410 Woodlands and Scrub.
- 420 Moorland and Heath.
- 430 Grassland.
- 440 Fen and Freshwater Marsh.
- 450 Aquatic Vegetation.
- 460 Saltmarsh, Dunes, Shingle.
- 470 Marine Vegetation.
- 480 Life-histories of Species in relation to environment.
- 490 Floristic Lists.

500 Animal Life (Zoology and Animal Ecology).
- 510 Terrestrial Surface Animals.
- 520 Aerial Animals.
- 530 Animals intimately associated with terrestrial plants.
- 540 Subterranean Animals.
- 550 Freshwater Aquatic Animals.
- 560 Marine Littoral Animals.
- 570 Marine Animals.
- 580 Life-histories of Species in relation to environment.
- 590 Lists of Fauna.

600 Prehistoric Man (Prehistory).
- 610 Pre-Palaeolithic.
- 620 Palaeolithic.
- 630 Neolithic.
- 640 Bronze Age.
- 650 Early Iron Age.
- 660 Prehistoric Earthworks, Megaliths, Roads, etc.

700 Historic Survey.
- 710 Roman Occupation.
- 720 Early English.
- 730 Norman.
- 740 Late Mediaeval.
- 750 Renaissance.

760 Post-Renaissance.
770 Industrial Period.
780 Period of the Great War and later.

800 Economic Survey.
   810 Population (number, distribution, anthropometry).
   820 Land (tenure, ownership, utilisation, value).
   830 Agriculture, Forestry, Fishery.
   840 Mining and Quarrying.
   850 Manufacturing Industries.
   860 Engineering and Building.
   870 Communications and Transport.
   880 Distributive Industries (markets, shops, etc.).
   890 Finance.

900 Social Survey.
   910 Occupations.
   920 Housing and Public Health.
   930 Government and Administration.
   940 Military and Naval Organisation.
   950 Education.
   960 Recreations (sports, games, etc.).
   970 Language and Culture.
   980 Ecclesiastical and Sectarian Organisation.
   990 Religion and Religious Influences.

CHAPTER III

THE SURVEY AREA OR "REGION"

IN a strict sense the term "Region" is perhaps misapplied to the small local areas usually chosen for intensive surveys of the kind with which we are here concerned. The term *regional*, in fact, is generally used in antithesis to *local*. For this reason the late Professor Herbertson objected to the use of the title "Regional Surveys" for what he would have preferred to call "District Surveys", but the former term has now become well established. Nor is its use entirely without justification; for it is the breadth of outlook in regional survey that distinguishes it from mere topographical description. It is true that our survey areas often do not merit the title of geographical regions, but even the smallest of them is set in such a region and it is in this regional setting that we wish to study it. We may thus achieve results that will form valuable contributions towards regional surveys in the larger sense.

The successful execution of a survey will depend in large measure upon the initial choice of the area to be surveyed and this is by no means a simple matter. Although we shall conclude the present chapter by recommending a particular method of delimiting local survey areas we have thought it desirable to explore all the possible methods because of the mistakes that have sometimes been made. There are many considerations, both theoretical and practical, that must be taken into account in choosing a survey area. From a practical point of view it should be small enough to admit of adequate treatment by a group of workers operating from a centre.

Let us think first of the larger problem, of which ours is a part, namely, that of dividing an extensive tract of country into small local areas; for one of the aims of the regional survey movement is to cover the whole country by locally organised surveys. There are three principles upon which a large area may be subdivided, namely, into

(1) Natural regions having boundaries determined by physiographical features, or social groupings.

(2) Areas having administrative boundaries for their limits.

(3) Areas having geometrical outlines.

Each of these has its advantages and its drawbacks for our purpose and we shall naturally wish to secure as far as possible the advantages of all.

The subdivision into natural regions is theoretically the most desirable but practically the mostly difficult. It might almost be said that the discovery of the natural region is one of the final achievements of a survey rather than its starting point. It is only by virtue of surveys already made for the purpose of preparing maps that we can even attempt such a subdivision. If our surveys were to be purely physiographical, the river basins would afford appropriate regions, except that most of them would need further subdivision; but watersheds do not always separate human groups from one another. When we view the problem of natural regions from the human standpoint we are faced by all the considerations set out under headings 800 and 900 in the conspectus. It will be seen that in most cases a region having its limits determined by such considerations can only be defined when the survey itself has reached an advanced stage. In rural districts the area which sends produce to the same market town or that in which a local newspaper circulates will often be found to delimit a human region. The more data of this kind that can be obtained by a survey organisation about to decide this important question the more likely is the area chosen to stand the test of time as the survey proceeds.

At the opposite extreme from the natural region is the geometrical area which may be either rectangular or circular. Except for the general fascination of circles and the fact that they may be described with compasses, thus saving much deliberation, there is little to be said in favour of a circular area. To have the survey area fairly evenly disposed about its centre is undoubtedly an advantage, but from this purely practical point of view accessibility is of greater importance than mere distance. It will be found that some places on the circumference of a circle are more difficult of access than others beyond it. In any case convenience, which will carry little weight with the genuine student, must not be made a first consideration. The regional surveyor will make it his business to visit and study the less accessible parts of his region and he will be amply repaid for his trouble.

On the other hand, the positive disadvantages of circular areas are such as in our opinion to rule them out entirely. In the first place, the compass pencil has a sublime disregard for all natural features. On the map it will mount lightly over hills, cross rivers and bisect village communities with complete indifference. For this reason some societies having commenced to survey a circular area have gradually modified it until it has lost all semblance of its original form. As subdivisions of a larger region circular areas are quite the least desirable. A series of circles on a map must either overlap very considerably or leave awkwardly-shaped areas uncovered or both. A certain amount of overlapping of adjacent survey areas is not in itself undesirable; on the contrary, if it promotes intercourse between neighbouring survey organisations, with comparison of results and mutual help and criticism, it is a positive gain. But the uncontrollable lenticular overlappings of circular areas are not the kind we should choose.

The case for rectangular areas is somewhat different. A rectangle is much more adjustable to local needs than a circle. There need be no overlapping of adjacent survey areas when a larger region is subdivided into rectangular blocks, or there may be just as much as we choose in each of the four directions. There are two ways of determining the boundaries of a rectangular survey area. We may draw a suitable rectangle around our centre of population, or we may divide a larger area into rectangular blocks by drawing two sets of parallel lines on the map at right angles to one another. One feature of the latter plan is that the whole area is covered without overlapping and it is the plan necessarily adopted by the Ordnance Survey for the purpose of its regular map issues. For this reason so long ago as 1896, Dr Hugh Robert Mill, in a paper read before the Royal Geographical Society, put forward a plea for a "Geographical Description of the British Islands based on the Ordnance Survey".[1] Dr Mill's proposal was that a separate geographical memoir should be prepared for each sheet of the map on the scale of one inch to a mile and he followed it in 1900 by a specimen memoir on South-west Sussex.[2]

---

[1] *Geographical Journal*, April, 1896.
[2] "A Fragment of the Geography of England", *Geographical Journal*, March and April, 1900.

These two papers, with the published discussions which followed them, stand as a landmark in the history of the regional survey movement. The memoir, prepared during great pressure of other work, fell far short of the author's ideal but it is worthy of careful study as a model by all who are interested in regional surveying. Dr Mill's proposals did not, however, meet with the direct response for which he had hoped and some of the participants in the discussion criticised the mode of division of the country into the arbitrary rectangular areas of the 1-inch Ordnance Survey sheets. These papers, however, have had a considerable influence in stimulating local geographical research. The preparation of sheet memoirs by students graduating for geography diplomas became popular for a time at the Oxford School of Geography, the London School of Economics and elsewhere and some of these students' theses have been published.[1] The scope of the proposed sheet memoirs cannot be said to encompass the ideals of a regional survey as we now understand it, but had they been prepared and published they would have afforded one of the most valuable sources of information for regional surveyors.

Neither the Ordnance Survey sheets as published nor any *externally determined* rectangular areas are well suited for local surveys. We have already noted that a civic centre (city, town or village) is the essential nucleus of a survey area. A single sheet of the published map will often include several important centres or, on the other hand, one town may occupy corners of as many as four Ordnance Survey sheets. Also, for intensive study, whether by a survey society or by a school, the 216 square miles included in even a small sheet is, as a rule, far too large an area.

[1] E.g. *The Reigate Sheet*, by Ellen Smith.

A rectangular area described about a civic centre has much to recommend it. We shall advise its adoption for the purpose of procuring base maps as described in Chapter IV, but before explaining the method of fixing the size and shape of the rectangle we will consider the case for survey areas coinciding with local administrative units. It has often been objected to administrative areas that they pay no heed to physiographical features. This is only partially true. On the other hand, historically and to a great extent at the present day, they have a very important human significance. We may be certain that their boundaries have been determined by regional factors even though we may not readily detect what those factors have been.

We shall gain freedom in the initial task of delimiting our region if we fix upon the smallest administrative unit, namely, the civil parish, as the *unit of area* for survey purposes. A local survey area will then consist of one or more of these units. In the case of a school survey it will usually be the single parish in which the school is situated. For a regional survey organisation a suitable area will be a town or city with a number of the surrounding civil parishes. Local knowledge will help the group to decide which parishes should be included. The size of the area will depend upon the nature and strength of the organisation undertaking the survey. Experience has shown that about 100 square miles is the maximum for a workable area for a local society. It is obviously better to achieve a fairly thorough survey of a small region than to make an abortive attempt at a too ambitious task. A local representative residing in each of the parishes included in the survey area, particularly if he is a schoolmaster directing an educational survey of the parish, will be a

THE SURVEY AREA OR "REGION"

very great asset to the central organisation. Experience has also brought home the fact independently to many who have engaged in regional survey work that the civil parish is the most desirable unit of area. Much of the already available material (e.g. census and agricultural returns) has been published on a parish basis. The civil parishes may, indeed, be likened to the cells within a living organism, each complete in itself with its nucleus and yet in vital contact and communication with the neighbouring cells. The town or city might then be regarded as a controlling nerve centre, though after a century of unplanned development it has often assumed too much the nature of a malignant growth sapping the life of its rustic environs.

A town with a number of its surrounding parishes outlined on a map will form an irregular figure like a group of parts of a jig-saw puzzle. While this will constitute the area for intensive study as described in the later chapters, there is no reason why it should occupy the whole of the base map upon which the results of the survey are to be summarised. There are in fact advantages to be gained by squaring it off and taking in an additional margin. To square up the chosen area it is only necessary to draw horizontal lines through its most northerly and southerly points and perpendicular lines through the extreme points on the east and west. The margin may then be added by drawing lines parallel to these at suitable distances thus determining the area of the base map. The diagram (Fig. 2) shows a hypothetical base map arrived at in this way in which A is the civic centre and the shaded portion its chosen surroundings. This has been squared off by the dotted lines and expanded to include a small marginal area. A similar course will be adopted

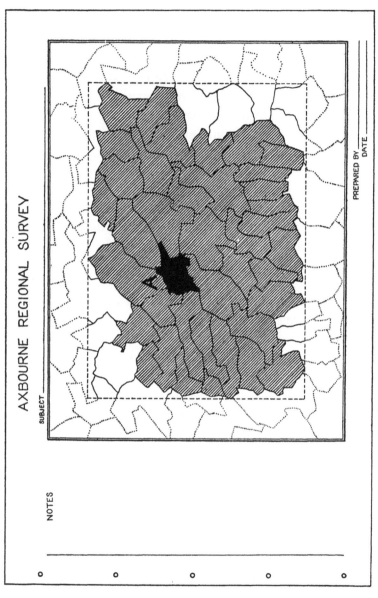

Fig. 2. A diagram illustrating the method of determining a small-scale base map.

when the survey area is a single parish but in this case the scale of the base map will be larger. An example of an area consisting of only one parish is shown in Fig. 4 on page 44. The outer edges of such a map will encroach upon the territory of neighbouring survey areas. But this, as we have already remarked, is an advantage; for it will remind us that our area is not an island and will necessitate a very desirable co-operation with neighbouring survey organisations if such are in being. We shall discuss the methods of dealing with the marginal portions of the base maps in Chapters IV and VIII.

Although the civil parish has great advantages as the survey unit, the various groupings of parishes for civil or ecclesiastical administration (e.g. rural districts, rural deaneries) are often unsuitable for survey areas. Nor would the advantages to be gained by adopting them be at all comparable with those attaching to the parish as a unit. The counties, however, as a basis for co-ordinating local surveys under a larger scheme as the survey movement progresses, have the same kind of advantages as the parishes. Not only are their boundaries of historical significance and their position in present-day administration of great importance, but an immense volume of regional research is already published about the separate counties. We have our county histories, floras, archaeological publications, directories and so on. The further grouping of the counties into "Provinces" or "regional divisions" has been much discussed and indeed carried out for various practical purposes during recent years. The whole subject has been fully and ably dealt with by Professor C. B. Fawcett in a work entitled *The Provinces of England*. We may thus foresee a time, perhaps not too far distant, when by the linking up of detailed parish surveys with local surveys, of these with county surveys and of these again with surveys of larger regions we may achieve Dr Mill's ideal of a "Geographical Description of the British Islands" and much more, but on a sounder regional basis than that of the Ordnance Survey sheets.

In considering the limits we should set to a local survey area our attention has perforce been directed also to its immediate surroundings. Having finally decided those limits and for the most part excluded the environs of the chosen area from the base map, we must by no means exclude them from our thoughts. On the contrary, we must again direct our attention to them broadly as a study collateral with the intensive investigation of our own small area, which we must endeavour to see from without as well as from within.

Our insight into our own region will be seriously limited if we concentrate our attention too exclusively upon it; for instance, its geological structure can only be comprehended as part of the geology of a wider region. We shall therefore need to study and if possible incorporate in our collection of survey material the 1-inch and $\frac{1}{4}$-inch geological maps which include the area of our base map and also make transects extending across and beyond our area in both directions.

If we wish to understand our local weather we shall need to take cognisance of the phenomena of the atmosphere over a still wider area, including not only the whole of the British Isles, but parts of the Continent and a wide margin of the Atlantic. We shall accordingly add to our survey collections small scale climatic charts of this wide region as well as maps on intermediate scales, increasing in detail and decreasing in area, so far as these can be obtained or prepared. On all such charts the position of our area should be indicated.

Again the streams which flow in our area may have their sources beyond it and they will leave it to join larger rivers and ultimately to reach the sea. The hydrographic maps of the region should be preceded in the atlas by a small scale map showing the whole of the drainage system to which it belongs.

When we come to deal with the human aspects of the region we shall need additional maps to show whence come its roads and railways and whither they go, how they link up with the national system of communications and so on. The "High Streets" which figure so largely in the town maps of, say, Croydon and Rochester appear as infinitesimal portions of the arterial roads to Brighton and Dover in a road map of England.

We need first of all to understand the geographical position of our region, to associate its villages and roads with more distant places and ultimately to form a conception of our civic centre in its historical and present-day relations to the whole country and to the world in general.

How can we most vividly depict the position of our town or village on the face of the earth? The most obvious way is to mark it by a dot on a map of the world, but this will neither help our imaginations nor add anything to our knowledge. We already have a map of the world in our mind's eye and the whole of the British Isles are scarcely more than a few dots upon it. A more vivid method of bringing out the relation of our town to the world is by means of an "Orientation Chart" with the heart of our region at its centre. Varied specimens of such charts may be seen set up at the Cabot Tower,

Bristol, on the North Hill, Great Malvern, on Hindhead, Surrey, on Dundee Law, at Stirling Castle and elsewhere.

A simple orientation chart can be prepared in the following manner. First describe three concentric circles of, say, $1\frac{1}{2}''$, $4''$, and $7\frac{1}{2}''$ radii. Within the innermost circle copy the black outline 1-inch Ordnance Survey map for $1\frac{1}{2}$ miles radius around the point for which the chart is to be constructed. Between this and the next circle, using the scale 1-inch = 20 miles, mark down the positions of towns, landmarks and places of special interest within a radius of $51\frac{1}{2}$ miles. Within this space may also be drawn in broken lines circles at 10-mile ($\frac{1}{2}$-inch) intervals. In the outer annular area more distant cities and the world's chief capitals should be placed in their correct positions as to direction but not, of course, as to scale. The scale of distance must decrease more and more rapidly as the actual distance increases but not in jumps that may make cities appear in the wrong order on the chart. Against each city, etc., its name and distance in miles should be written. Radial lines may be drawn from the 4-inch circle to the cities shown in the outer ring. The chart will be completed by indicating the points of the compass around the circumference of the outer circle.

In a school the preparation of a chart of this kind affords an excellent exercise in both geometry and geography. The directions and distances of remote places cannot, of course, be measured on a map of the world. They may either be calculated as exercises in spherical trigonometry from their latitudes and longitudes or measured on a terrestrial globe. If possible a copy of the chart should be engraved upon metal or mounted in such a way as to protect it from the elements, and set up in correct position at a point of vantage such as the school roof or the church tower, if these are accessible, or on a pedestal on a high open space. There is something peculiarly stimulating to the imagination in the contemplation of one of these charts. If we can stand upon an eminence and say "Athens is straight over there" or "Cape Town is precisely in that direction" these places with all their associations immediately become more real to us. We gain a vivid impression of the fact that they do exist on this same earth and have a geographical relationship to the intimate details of our own town or village as depicted in the centre of the chart.

CHAPTER IV
MAPS

"The natural development of the map is due to the desire which necessity or curiosity imposes upon mankind to explore the earth's surface, and to move from one point on that surface to another, working from the known to the unknown along the path of experience and enquiry."

SIR H. G. FORDHAM in *Maps, their History, Characteristics and Uses.*

HAVING decided the geographical boundaries of the region the next step will be to obtain a supply of outline base maps upon which to record by colours, symbols or lettering the data collected as the survey work proceeds. It is surprising to what extent regional observations lend themselves to graphic representation upon maps, but the beginner should guard against the tendency to crowd too much into a single map. The uses to which the maps will be put will be more fully discussed in Chapter VIII. The present chapter we shall devote to considerations affecting the selection of a map suitable for the purpose of a base map. We shall first give a specification for an ideal base map and then discuss the various possibilities of approaching more or less nearly to this ideal.

1. The map should be of such a scale that it can be printed on a sheet of convenient size. If it is to be reproduced from a manuscript drawing any scale which fulfils this requirement may be chosen. If, however, a reproduction of part of a published map is desired the selection will be more limited. When the survey area consists of one or two parishes the Ordnance Survey map on the scale of 6 inches to 1 mile is the most suitable basis for the outline map, but if it comprises a civic centre with

an extensive rural environment the scale of 1 inch to 1 mile should be adopted. For areas of 100 square miles or more base maps on the scale of ½ inch to 1 mile may be used with success.

2. The maps should show the coast-line, if any, river and stream courses, contour lines and, if possible, the boundaries of geological outcrops. While a set of maps will be utilised to depict the physical features of the region it is desirable that these should be present though not obtrusive on all copies of the base map.

3. Without undue crowding the base map should embody sufficient topographical detail, i.e. roads, railways, village sites, buildings, etc., to enable accurate location of any point to be made. The boundaries of civil parishes should be shown.

4. In the case of a small scale map when the detail is necessarily condensed it is desirable to have the maps printed in grey or brown rather than black so as not to obscure the symbols or other matter to be added in manuscript.

5. The sheet should allow space for subject titles and ample room for marginal notes, including an index to the colours or symbols used.

6. The maps should be procurable in large quantities at a reasonable cost so that they may be freely used for rough work and for experimenting in methods of representation, as well as for the finished maps to be placed in the survey portfolio. For the latter purpose it is highly desirable that a proportion of the supply of base maps should be printed on paper of superior quality.

In cases where the cost of purchasing a supply of base maps (e.g. out of the too meagre grants to elementary schools) constitutes an obstacle to the commencement of a survey, maps made by a hektograph or "Plex" duplicator may be used with success though this course cannot be regarded as ideal. On the other hand, some regional survey workers prefer to dispense with base maps altogether and to prepare a manuscript map for each subject, inserting only such detail as is required.

We will deal first with the various published maps that may be adopted or adapted for use as base maps. First and foremost amongst these are the Ordnance Survey map issues. The Ordnance Survey Dept. offers special facilities to *bona fide* regional survey societies for procuring supplies of base maps. Not only is the privilege of the scale of special prices for educational maps extended to such societies but the Department will give quotations, based upon that scale, for maps specially printed to meet the requirements of a local survey as discussed below.

If a society elects to make its own maps, either by some process of duplication or by printing from a special block, it will in every case be necessary first to obtain the permission of the Controller of H.M. Stationery Office in whom the copyright of Ordnance Survey maps is vested. The form of application for such permission is obtainable from the Ordnance Survey Office, Southampton.

The scale of special prices for educational maps in quantities of 50, 100, 200, or 500 copies is set out in the forms of application for them (O.S. 318 for Ordnance Survey maps and O.S. 46 for Geological Survey maps). These forms may also be obtained from the Ordnance Survey Office.

Among the maps so supplied the following list includes all those which are suitable for base maps for regional surveys. The dimensions given in brackets are those of the map sheets.

MAPS

Scale of 1 inch *to* 1 mile

1. The Outline Edition of the Popular Map (27"×18").
2. The Black Outline Edition (18" × 12").
3. Uncoloured Geological Outline Maps (18" × 12").

Scale of ½ inch *to* 1 mile

4. Half-inch Layered Map (27" × 18").

Scale of 6 inches *to* 1 mile

5. Printed in black with contours (18" × 12").
6. Ditto with geological boundaries and symbols (18" × 12").

Let us now review the merits of these several issues as base maps for the various types of survey area discussed in the preceding chapter. We shall not enter into the question of relative cost. The difference in the initial outlay for one or other of the suggested maps is comparatively small and our view is that a society undertaking a regional survey should aim at procuring the base map that is best suited to its purpose.[1] This course will pay in the long run. It may be noted, however, that the larger the number of maps required the less the cost per copy. A society will be ill-advised to obtain less than 200 copies of the base map.

It will rarely happen that an Ordnance Survey sheet as published will be suitable for the base map. More often parts of one or more sheets specially printed as a single map will be required, and it will be necessary to obtain a quotation from the Ordnance Survey Dept.

1. *The Outline Edition of the Popular Map.* This map has advantages over all other small scale maps in that the

[1] Societies are invited to seek the advice of the authors, c/o The Hon. Secretary, Regional Survey Committee, The Geographical Association, Aberystwyth.

contours are inserted at 50-foot intervals and that the topographical detail is more up-to-date. Its chief drawbacks from our point of view are the omission of the parish boundaries[1] and the absence of geological boundaries. The contours are printed in red and the rivers in blue. If this issue is adopted it will be advisable to have the contours printed in grey instead of red in order to keep them subdued. It is an advantage rather than otherwise to have the water coloured blue on the base map. The sheets as published cover an area of nearly 500 square miles, which is very much in excess of a workable survey area. The sheets also (27" × 18") are much too large to form convenient base maps. There is, however, a fair chance that the chosen area will fall wholly within one of the published sheets; otherwise it will be necessary to have a base map made up of parts of two or more sheets.

2. *The Black Outline Edition.* This map, as its title indicates, is printed wholly in black. As a base map it should, as already stated, be printed in grey. It has the advantage over the Popular edition of showing the parish boundaries, a feature of some importance from our point of view. The contours, however, are given only for 100-foot intervals and the topographical detail is less up-to-date. Notwithstanding these disadvantages we prefer it on the whole to the Popular edition for the purpose of a base map. The sheets as published (18" × 12") cover 216 square miles but if the chosen area falls wholly within one of them it will not be too large for a base map. The portions outside the area will form the margin mentioned in the previous chapter.

[1] The parish boundaries are, however, to be re-inserted in the next revision of the 1-inch map.

3. *Uncoloured Geological Outline Maps.* These are the same as the preceding with the geological boundaries and symbols added. For areas in which these maps are available they are without question the most suitable for small scale base maps, and they have the added advantage of being the cheapest of the educational issues. Unfortunately there are still large areas of the country for which the Geological Survey maps have not yet been published in this series.[1] For these areas uncoloured copies of the old series of geological maps are obtainable, but while one copy is essential to the survey they are not so desirable as base maps. Fig. 3 is a specimen of the uncoloured geological outline map (new series) and Fig. 3 a a horizontal section across it.

The geological boundaries are so desirable on the base map, or at least on a number of copies of it, that the possibility of their inclusion should always be explored. If otherwise suitable maps showing them cannot be supplied by the Ordnance Survey Office they should be compiled from the best available sources and either added in manuscript to a number of copies or overprinted by the "true scale" process described below.

For many parts of the country where the geological survey has not been revised only the original geological maps are obtainable. These are hand-coloured and done on the earliest 1-inch ordnance maps on sheets measuring $35\frac{1}{4}'' \times 23''$ or in some cases quarter sheets. The hand-coloured maps are expensive but uncoloured copies can be obtained direct from the Ordnance Survey Office at 2s. 6d. per sheet or 1s. per quarter sheet. These are not suitable for base maps.

[1] See Catalogue of Geological Maps, Memoirs, etc. published for H.M. Geological Survey by the Ordnance Survey Dept., 1s. net.

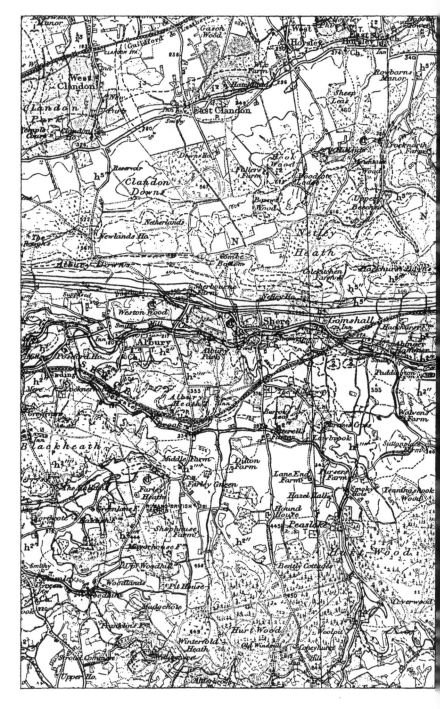

Fig. 3. A specimen of the modern Uncoloured Geological Outline Map.

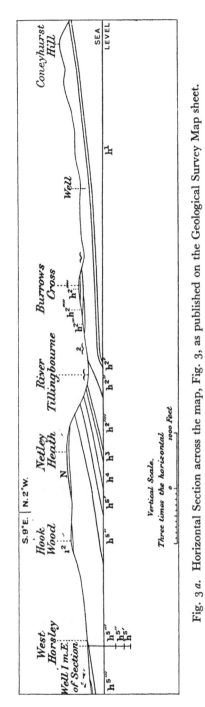

Fig. 3 a. Horizontal Section across the map, Fig. 3, as published on the Geological Survey Map sheet.

4. *Half-inch Layered Map.* When the chosen area is a large one, say 100 square miles or over, the survey society may prefer a base map on the scale of ½ inch to 1 mile. Most of the maps to be prepared will lose little by the reduction in scale while the small size of the sheets is a distinct advantage for filing purposes. The coloured "layers" which appear upon the map as published will of course be omitted from the base maps. The contours at 100-foot intervals are in red and the water in blue on the published maps. These may be dealt with as suggested in the similar case of the Outline Edition of the Popular Map (see above). This map does not show either the parishes or the geological boundaries but at an extra cost either or both of these features could be added. The only real advantage in a ½-inch base map is the smallness of the sheets and the area of paper to be coloured in cases where the survey area is a very large one.

5. *Six-inch maps printed in black with contours.* When the chosen area is a small one, e.g. one or two civil parishes, the base map will of course be on a correspondingly larger scale. It may be either the 6-inch Ordnance Survey map or a reduced copy of this. The 6-inch map is now issued in quarter sheets (18″ × 12″) each representing an area 3 miles by 2 miles. It very rarely happens that the whole of a parish falls upon one quarter sheet and if the 6-inch map is to be used as a base map in its published form it will usually be necessary to mount two or more quarter sheets together. Apart from the inconvenience of doing this the cost will be multiplied by the number of quarter sheets involved and may easily become prohibitive. In such cases it will be much cheaper to have a special map prepared by the Ordnance Survey Dept. This may either be a reproduction of the 6-inch map to

scale or a reduction or enlargement of it. If the 6-inch scale would necessitate an inconveniently large base map it may be reduced to, say, the scale of 4 inches to 1 mile. Maps on a reduced scale are correspondingly less expensive but it must be borne in mind that the quality of a map is seriously impaired by too great reduction. The lettering and "characteristics" in Ordnance Survey maps are designed to give the maximum legibility and best cartographic effect for the respective scales for which they are used. Photographic reduction as used in making plates if carried too far renders some of the lines too thin for clear printing and the whole map too condensed for comfortable reading. For this reason the reduction of 6-inch maps to scales of less than 4 inches to 1 mile is not recommended.

6. *Six-inch maps with geological boundaries and symbols added.* For areas in which these maps are available they are, of course, preferable to the foregoing. The same comments apply to them as in the case of number 3 above.

The Ordnance Survey Dept. will undertake to produce almost anything that is required in the way of a base map if the society is prepared to meet the cost and if the work of the Department permits. They will also supply to *bona fide* regional survey societies assortments of maps as published at rates below the retail prices.

A method of obtaining base maps which has many advantages and which will appeal to those having an aptitude for draughtsmanship is to prepare a drawing for reproduction by means of a printer's block. A reduced facsimile of a map made in this way is shown in Fig. 4 on p. 44. It will, of course, be necessary to obtain the permission of the Ordnance Survey Dept. for the compilation of such a map and a small royalty may be

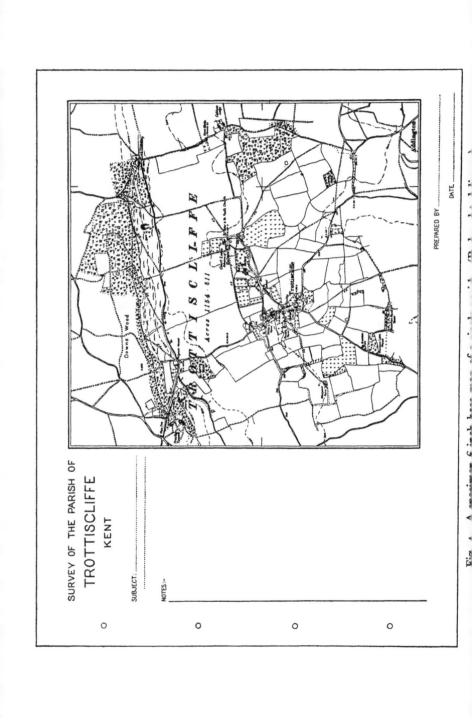

Fig. 1. A specimen 6 inch bus map of a single parish (Trottiscliffe).

charged. By adopting this method a survey organisation will have complete freedom of choice in the matter to be included and in regard to scale. The map can be compiled from various sources. For instance, the 50-foot contours and recent topographical changes can be taken from the Popular edition and still more recent developments may be sketched in if desired, the geological lines may be compiled from the best available sources and so on. If no professional cost is involved in preparing the copy for the blockmaker this method will not be much more expensive than any other in the long run. The block will be the property of the survey society and the maps may be printed locally as required on any paper (e.g. water-colour paper, transparent paper). It is desirable to make the original one and a half to two times the scale of the required base map. In this way the bold outline and easier lettering are reproduced with a degree of fineness that cannot be obtained by amateurs with the pen on a small scale. As an example let us suppose we are preparing a base map on the scale of $1\frac{1}{2}$ inches to 1 mile and that our copy is the 1-inch Ordnance Survey map. We shall first square off the required area on the copy and then divide it into $\frac{1}{2}$-inch squares by fine pencil lines. We shall next set out on our drawing paper a similar rectangle on a larger scale, say, $2\frac{1}{2}$ inches to 1 mile, and divide it by fine pencil lines into squares of $1\frac{1}{4}$-inch side. We may then copy in pencil the roads, contour lines and any other detail required square by square. If the map we have chosen for our copy does not bear the geological lines, parish boundaries or any other desired data we shall set out the squares to appropriate scale on other maps in order that we may incorporate these features in the manuscript. Before inking in the lines with drawing ink such lettering as is required should be inserted so that small inconspicuous breaks may be left where necessary in the line work. The freehand method of drawing within the squares should be sufficiently accurate for our purpose, but the use of proportional compasses will facilitate the filling-in of the detail besides increasing the accuracy. The foregoing instructions will be superfluous to those who are capable of preparing a good base map, but we have thought it desirable to give them because in the course of the survey work it will frequently be necessary for less experienced draughtsmen to essay the making of manuscript maps and plans. The same principles are involved, for instance, in enlarging portions of the 6-inch map as suggested in Chapter VII (see Figs. 7 and 8).

It will sometimes happen that a limited number of copies of a map, an enlarged plan or a diagram are required in connection with regional survey work. In such cases it may be economical to resort to what is known as the "true scale" process, which is now largely used by engineers and architects in preference to the older blue print method of reproduction. Several of the firms who cater for draughtsmen now have the plant for this process, often giving it their own trade name. True-scale prints are best made from tracings or drawings in Indian ink on transparent paper or linen, but this is not essential. The prints can be made in any colour and on any kind of paper to order. Their price varies with the size of the sheet. Small drawings are relatively more expensive than larger ones and in all cases the first copy costs about twice as much as subsequent copies of the same order. For prints of about 5 square feet in area on ordinary thin drawing paper the cost is about 4d. per square foot per copy. The remarks made above regarding copyright apply to maps made in this way.

Our interest in published maps is by no means exhausted when we have selected a base map and procured the required number of copies (say 200 to 500). As the work of the survey proceeds we shall find it necessary to refer to published maps of all kinds from the sixteenth-century county maps of Christopher Saxton to the most recent productions of the Ordnance Survey Dept. Maps, old and new, are in themselves an important source of information concerning any region and their production has necessarily been preceded by a vast amount of survey work of a highly technical character. It should be the aim of a survey organisation to acquire as a part of its survey collection as many as possible of the maps upon which the chosen area is represented either wholly or in part. Some of the older maps are scarce and valuable but they can usually be seen and studied in a library and if possible photographic or manuscript copies should be made for the survey collection. The beautifully engraved early editions of the 1-inch and 6-inch Ordnance Survey maps are specially valuable for reference. These are now called "Record Maps" by the Ordnance Survey Office and are still supplied as such to order. They are expensive because they have to be specially printed from the old plates. Actually the very earliest edition cannot be printed from the old plates as those plates were altered by adding later detail. Prints from the plates so altered can be supplied, as also spare copies of the first edition that still remain in stock at Southampton. It has recently been decided by the Ordnance Survey Dept. to make photographic reproductions of those of which no more copies are left, as demand arises. These will be sold at 5s. each.

We need scarcely add that those interested in regional survey should never leave copies of old Ordnance Survey maps in the possession of second-hand booksellers.

The principal modern issues of Ordnance Survey maps have already been mentioned but there are several others that may be used with advantage by regional surveyors. Among these are the index maps. One of the most useful is the ½-inch County Administrative map. In addition to the ordinary topographical detail of the ½-inch maps these show the parish boundaries, boroughs, urban and rural districts and parliamentary divisions. Each county is published on a separate sheet at 4s. 6d. net. The map for the county of London is on the 1-inch scale.[1]

County index diagrams on the ¼-inch scale are also issued. These are of two kinds sold at 6d. and 2d. respectively. The first has the sheet lines of the 6-inch and 25-inch maps overprinted in red on the ordinary ¼-inch map. The 2d. sheet gives no topographical detail but merely shows in black outline the boundaries of civil parishes within the county and the sheet lines. In its older form the index to the small sheet series of 1-inch maps is also included.

If the base map is on a small scale it will be necessary to acquire at least one set of 6-inch maps covering the whole area. It is indeed desirable to have several such sets. For the surface utilisation survey described in Chapter VI two sets (one for use in the field and one for preparing coloured copies for the portfolio) are essential, and for several of the subjects mentioned in Chapter IX (e.g. field names, rights of way), 6-inch maps

[1] As we go to press we learn that the style of these maps will be altered as soon as the Poor Law Unions have been abolished. The new style will be in three colours and clearer.

will be needed. These can only be procured at the cheap educational rates when a number of copies of a single sheet are required. As already mentioned, however, assorted sheets can be obtained direct from the Ordnance Survey Office at a discount by a *bona fide* regional survey society.

The 6-inch maps are almost indispensable to field workers in any branch and each active member of a survey organisation will probably wish to equip himself with a set. It should not therefore be difficult to make up a substantial order. In the case of a school survey of a small area the base map itself will probably be on the 6-inch scale and this difficulty will not arise.

Field workers have various ways of dissecting maps for convenient use in the field. The authors find it most convenient to cut up all maps, of whatever scale, into sections measuring 9" × 6". All the regular Ordnance Survey map issues are capable of exact subdivision in this way. The 6-inch quarter sheets measure 18" × 12" and will thus, when the margin is cut away, be divisible into four small field sections. Each of these is mounted on thin card $9\frac{1}{2}$" × $6\frac{1}{2}$", and marked in the top right-hand corner with an appropriate number. Each section will correspond exactly with a 25-inch scale map sheet and the numbers applied to these sheets may, therefore, be very conveniently applied to the field sections of the 6-inch map. A sheet of the popular 1-inch edition will cut up into nine sections of this size and one of the small sheet series as used for the Geological Survey into four. The Ordnance Survey Dept. will cut up a proportion of the maps in this way by machine if requested to do so. They will also mount the sections in covers if required, but this, of course, will entail additional cost.

The reader is strongly recommended to acquaint himself with the following catalogues and books dealing with the subject of this chapter.

ORDNANCE SURVEY. *A Description of the Ordnance Survey Large Scale Maps.* Ordnance Survey Office, Southampton. 6*d*.
—— *A Description of the Ordnance Survey Small Scale Maps.* Ordnance Survey Office, Southampton. 6*d*.
—— *Catalogue of Maps and Other Publications of the Ordnance Survey.* Ordnance Survey Office, Southampton. 1*s*.
GEOLOGICAL SURVEY. *List of Memoirs, Maps, Sections, etc.* Published by the Geological Survey of Great Britain. H.M. Stationery Office and Ordnance Survey Office. 1*s*.
FORDHAM, H. G. *Hand List of Catalogues and Works of Reference relating to Carto-Bibliography and Kindred Subjects for Great Britain and Ireland.* Cambridge University Press. 2*s*. 6*d*.
—— *Maps: Their History, Characteristics and Uses.* Cambridge University Press. 6*s*.
HINKS, A. R. *Maps and Survey.* Cambridge University Press. 12*s*. 6*d*.
LABORDE, E. D. *Popular Map Reading.* Cambridge University Press. 6*s*. School edition, 4*s*. 6*d*.

CHAPTER V

THE SURFACE UTILISATION SURVEY

WE cannot too strongly emphasise the value and economy of commencing a regional survey by making what is called a *surface utilisation* survey of the region. Its advantages will be more apparent when we have explained the method and will become increasingly so in the later chapters. In choosing our survey area we have staked out a claim, so to speak, and we wish to find out and record all there is to be known about it. The process, we need hardly say, never ends, and this is one of the differences between a published volume and a regional survey. The one is a closed book, and the other an ever-expanding collection of records and interpretations of regional investigations. The regional surveyor has indeed a voracious appetite for the books that have been written about his region, but he does not respect their bindings. In effect, if not literally, he will tear off their covers and re-arrange their pages with those of other books, and with gleanings from any and every source. But above all he should aim at acquiring an intimacy with every part of the region itself. Only by so doing will he be able to appreciate the value and the limitations of the work of those who have gone before. The surface utilisation survey is the first step towards gaining this intimacy.

The method is actually to look at every patch within the survey area and mark down its description upon the 6-inch maps, a set of which will be required for the purpose. In the case of a small survey area the base map itself will be on the 6-inch scale. The scheme of categories described in this chapter is adapted more particularly to predominantly rural areas, but the method is equally applicable to areas that are wholly urban. The 6-inch survey we are about to describe may be regarded as a preliminary stock-taking of all there is in the chosen area to be more thoroughly investigated in the years that follow.

Two important questions of technique have to be settled at the outset, namely, what categories of surface utilisation are to be differentiated in making the survey and what symbols or colours are to be used in representing them on the map. Let us first consider the categories into which we propose to sort out the various ways in which nature and man are utilising the surface of our region. In this connection two points should be borne in mind. The results of even the preliminary survey will be of some permanent value as records if they are dated. The work should therefore be divided amongst a sufficient number of surveyors to ensure its completion within one year. For this reason also the detail required must not be so great as to need more than a cursory examination of any part of the area. Again, the chosen categories must be such that they may be readily distinguished in the field by a person of average intelligence without a very special knowledge of any subject. The categories we have adopted as set out in Fig. 5 have been carefully chosen with a view to giving the maximum of usefulness compatible with these practical considerations. They have been tried out in several surveys and found on the whole to be very satisfactory. Whenever it is possible to record more detail on the maps than the scheme requires, without jeopardising the completion of the work within the year, this may be done with advantage. In fact any

THE SURFACE UTILISATION SURVEY 53

observation the surveyor is able to record in passing will be helpful in the later stages of the survey.

In representing the different categories on the map we may have recourse to symbols or colours or both. To some extent the surface utilisation is already indicated on the 6-inch ordnance maps by words or symbols. These data must not be accepted without verification as changes may have taken place since the map was made. A wood, for instance, may have been felled or a footpath lost. The symbols used by the Ordnance Survey are explained on the Characteristic Sheet issued by that Department and so far as they go we shall, of course, adopt them. But we shall also use a scheme of colours and other indications in the form of initial letters. It has been maintained by some that black and white symbols are preferable to colours because of the cost of printing maps in colour. It is not, however, anticipated that surface utilisation surveys will be published in the form of maps even in black and white and there is therefore no reason to forego the manifest advantages of the use of colours.

In working out the colour scheme we have taken the view that while colours are very useful in the close scrutiny of a map their chief value in cartography is to aid visualisation of the general features of the area depicted. For this reason a colour scheme should not be overloaded, as some have been, so that the finished map looks like a crazy patchwork quilt. When, in the preliminary survey, it is possible to record some degree of detail, as for instance the crops or the dominant trees in the woodlands, it will be better to indicate these by lettering than to attempt to differentiate them by separate colours. This may well be reserved for the vegetation and crop maps which will have independent colour schemes. Again, in the interests of visualisation, allied categories should be represented by different shades of the same colour, and should as far as possible suggest what they represent. Thus we have chosen shades of green to represent all the essentially rural categories, e.g. grassland, arable land, woodland and heath. The use of brown for arable has been advocated by some surveyors but in our opinion this interferes with the visualisation of rural as distinct from urban areas, whatever colour may be chosen for the latter. The yellowish green we have adopted for meadow land with its buttercups sufficiently differentiates this from the arable for which we have chosen a colder green suggestive of young cereal grasses and root crops.

We have chosen browns for the dominant urban categories and for roads, using a broken line for rights of way and a brown stipple for public open spaces. Thus we may think of the urban population issuing forth in bulk along the roads, trailing over the footpaths and spreading themselves out in dots upon the grassy commons and heaths.

Blue is the colour universally adopted in cartography for water. The more brilliant colours, vermilion and scarlet, have been reserved for categories, such as quarries and railways, which occupy only a small area on the map, so that the eye may readily pick them out.

A complete list of the suggested categories and the colours applied to them is given in Fig. 5, pp. 56, 57.

The working out of the colour scheme is only half the problem of the colouring of the maps. It remains to provide colours of standard quality that are easily obtainable. These must be transparent and fast and in a medium that will give good results even in the hands of an amateur

THE SURFACE UTILISATION SURVEY 55

colourist and on the kinds of paper upon which the cheap educational issues of maps are printed. Those who engage in regional survey work vary from town planners with a staff of trained draughtsmen to primary school pupils and boy scouts. To the former the colouring of maps will present no difficulty. They will be skilled in the use of water-colours and will take care to use paper that will take flat washes of these, the question of cost being a minor consideration. But to the vast army of amateurs the successful colouring of maps is more difficult. We have been very fortunate in securing the co-operation in this matter of Messrs Winsor and Newton, who have produced to our specification after many experiments a set of "Regional Survey Colours" which meet all these needs and are inexpensive. They can be obtained from Winsor and Newton, Rathbone Place, London, W., or from any of their agents. The range is light brown, middle brown, dark brown, buff, yellow, light green, middle green, dark green, olive green, orange, blue, violet, red, crimson and vermilion. These are the colours needed for the surface utilisation colour scheme, but the range will be found to be sufficient for colouring any regional survey map.

In the table in Fig. 5 the title of each category is given. An initial letter or other indication should be clearly written in pencil on the field maps as each area or object is identified by the surveyor. When a category represented by one colour is subdivided, as in the case of woodland, these letters will be transferred in Indian ink to the portfolio copies of the maps. There is no objection to doing this in all cases but the colours will as a rule be sufficient without them.

The quarter sheets of the 6-inch Ordnance Survey maps are inconveniently large for use in the field and we advise

SURFACE UTILISATION SURVEY

☐ Houses and gardens (not exceeding 3 acres) and any other buildings and their grounds not included below: LIGHT BROWN

☐ Schools, colleges, and their grounds (except playing fields): ORANGE

☐ Public roads and private roads used by the public: MIDDLE BROWN

☐ Churches, chapels, convents, vicarages, church halls, etc.: VIOLET

☐ Footpaths and bridle roads used by the public: MIDDLE BROWN, broken line

☐ Institutions (hospitals, almshouses, prisons, etc.): YELLOW, two applications

☐ Railways, railway stations, sidings, goods yards, etc.: RED

☐ Naval and military institutions: YELLOW, two applications with RED stipple

☐ Factories, warehouses, transport depôts, wharves, etc.: VERMILION

☐ Farms, including farmhouses, attached cottages, farmyards and buildings and farm roads: YELLOW

☐ Mines, quarries, exposed geological sections: CRIMSON

☐ Permanent grassland: meadow, pasture or grassland: LIGHT GREEN

☐ Waterworks, reservoirs: BLUE, single hatching

☐ Arable land: DARK GREEN

☐ Sewage disposal works: BLUE, cross hatching

☐ Market gardens (M.G.), kitchen gardens (K.G.), nurseries (N.), allotments (Al.): DARK GREEN, cross hatching

Fig. 5. Index to surface-utilisation

SCHEME OF CATEGORIES AND COLOURS

Plantations of small-fruit (giving initial of kind): DARK GREEN, single hatching

Orchards (giving initial of kind): MIDDLE GREEN, hatching on appropriate base colour (LIGHT or DARK GREEN)

Poultry farms, pig farms, etc.: YELLOW, single hatching

Woodland (giving initial of kind): MIDDLE GREEN

Scrub: LIGHT GREEN with O.S. symbol

Moorland and heath: OLIVE GREEN. Bogs indicated by the addition of BLUE stipple

Marshes: LIGHT GREEN with BLUE stipple

Sand, shingle, bare rocks, etc.: BUFF

Common land, public parks and recreation grounds: MIDDLE BROWN stipple on appropriate base colour (LIGHT GREEN, MIDDLE GREEN, etc.)

Public grounds with games (Cricket, Cr., Football, F., etc.): MIDDLE BROWN and ORANGE mixed stipple on LIGHT GREEN

Private sports grounds (Cricket, Cr., Golf, G., etc.), racecourses, school playing fields: ORANGE stipple on LIGHT GREEN

Burial grounds, other than churchyards: VIOLET, single hatching

Waste land, refuse dumps, slag heaps, etc.: MIDDLE BROWN, crosses

Lakes, ponds, rivers, streams, canals, etc.: BLUE

Seasonal streams, dried water courses: BLUE, broken line

Sea and large sheets of water: BLUE, margin merging into a pale wash of Prussian Blue water-colour

categories and colour scheme.

dealing with them in the manner described on p. 49. Equipped with the maps and having assimilated the table of categories the surveyor is ready to commence work in the field. It should be borne in mind that this preliminary surface utilisation survey is primarily a rustic one. Not only is the 6-inch scale too small to carry much differentiation in towns where the surface is cut up into very small areas, but the task of surveying them minutely would occupy too much time during the first year. We shall briefly discuss the survey of urban areas in the next chapter. For the present we shall content ourselves with the few urban categories indicated in the table. These, together with the roads, railways, churchyards and schools which are already shown on the map, may with advantage be coloured on the field maps before they are taken into the field. In this way the surveyor will save much time and will more readily be able to pick out the small areas in towns which need surveying. For the same reason it is advisable to colour on the field maps the portion surveyed after every excursion. It is not, of course, necessary to bestow the same care upon them as will be given to colouring the portfolio copies.

From a point of vantage such as a hill, a railway bridge or a church tower it is often possible to note the utilisation of fields, etc., over a wide area. So long as nothing is recorded without certainty there is no objection to this being done in the preliminary survey, especially if the chosen area is extensive. Field glasses are a great asset in this, but whenever fields are surveyed in this way they should be verified at closer quarters if opportunity arises.

When an item is found to be already correctly described upon the map (e.g. Wood, Nursery, Cricket Ground, F.P.) the surveyor will tick the description in pencil. If it is incorrectly described or not described he will make the necessary alteration and insert the appropriate initial letter from the table. Footpaths are not invariably marked upon the maps. When they are they usually, but not always, indicate rights of way. No part of a footpath or bridle road should be coloured on the map until it has been observed to exist as such.

Additional notes beyond those required for making the surface utilisation maps should always be recorded on the field maps when this will not involve a loss of time. The following suggestions may be borne in mind.

Permanent Grassland. It will be useful to distinguish between meadow and pasture. When animals are grazing in the field the fact may be recorded by writing "Cows", "Sheep", or as the case may be. If any species of flowering plant, e.g. oxeye daisy, meadow sweet, is very conspicuous in a field the fact may with advantage be noted. The presence of many rushes should always be recorded. If grassland is used for any other purpose than agriculture (e.g. cricket, football, golf) it must be noted.

Arable land. Special crops (e.g. hops) which do not come under the market-garden group should be recorded. Ordinary crops need not be specified though any notes that can be made regarding them or the soils will not be without value in later stages of the survey.

Small Fruit. The kinds should be specified (e.g. strawberries, Str.; raspberries, Rasp.).

Woodland. This is already indicated by symbols on the map. The existence of each piece of woodland should be verified and as far as possible the types of woodland, that is the dominant trees, should be recorded (e.g. beech, B.; birch, Bi.; oak, O.; pine, P.; larch, L.; ash, A.; coppice-with-standards, C.W.S.).

Scrub. This category is intermediate in character and often in succession between grassland and woodland. It varies from a few small hawthorns, briars or brambles commencing to invade grassland to dense thicket without woodland trees. When the process of invasion is in its early stages the field may be classed as grassland, but a note should be added on the map.

Water. Ponds, etc., are usually indicated on the ordnance maps by a heavy line on their north and west sides. When water is flowing in a course shown on the map the direction of the flow should be indicated by an arrow or the arrows, if any, on the map ticked. If a pond or watercourse shown on the map is dry the fact should be noted.

Quarries, "*pits*", *etc.* When these are shown on the map write "n.s." if no geological section is to be seen. When a quarry is not shown it should be sketched in, taking care not to exaggerate its size. In all cases where a section can be seen, however small, state whether clay, Cl.; sand, Sd.; gravel, Gr.; chalk, Ch.; etc. State also whether working or disused.

Factories, etc. State what these are. If they are temporarily or permanently out of use this should be noted. Brickfields shown on the map are sometimes in this condition.

Except where building operations have recently been in progress each area shown on the map will usually be occupied entirely by one of the categories in the table. When this is not the case (e.g. a field partly in grass and partly under crop) the subdivision can be sketched in by eye. All alterations in the map should be sketched in, but where owing to building or other operations these are considerable, a plan of the altered area can often be seen locally (e.g. in the possession of an estate agent) and the alterations copied from it.

When the survey is completed and the fair copies of the maps are coloured they will constitute a record that will increase in interest as time goes on. With what interest should we study such maps now if they had been made a hundred, fifty or even twenty years ago. It is not intended, however, that the maps shall be merely filed away as records, but rather that they shall furnish starting points for the more detailed study of the region. During the winter following the field survey the portfolio copies may be coloured at leisure and a statistical abstract of acreages, within the limits of the classification, may be prepared for the whole area and for each parish within it. The areas in acres to three places of decimals are given on the 25-inch Ordnance Survey maps. If access can be had to these maps the figures should be copied on the 6-inch maps. Failing this the fields may be measured with a planimeter or to a sufficient degree of approximation by the use of a piece of tracing paper ruled into acre and $\frac{1}{4}$-acre squares. The diagram (Fig. 6) is a part of such a transparency prepared for use with 6-inch maps.[1]

[1] An "acre grid" may be traced from Fig. 6 which is accurate for 6-inch scale maps or it may be constructed as follows:

First calculate the size of an acre square to the scale of 6 inches to 1 mile.

1 acre = 4840 square yards.

$\therefore$ 1 acre square measures $\sqrt{4840}$ yards = 69·57 yards.

1760 yards is represented by 6 inches.

$\therefore$ 69·57 yards = $\dfrac{6 \times 69\cdot57}{1760}$ = 0·2371 inches.

$\therefore$ 10 acre squares measure 2·37 inches

or 20 acre squares measure 4·74 inches.

Hence a square of 4·74 or $4\frac{3}{4}$ inches side can be divided into 400 acre squares on the 6-inch scale.

The large squares are acres. The field upon which it is imagined to be superposed will be seen to be 26½ acres in extent if the squares are counted up. A little practice will enable one to manipulate the transparency so that as few part squares as possible occur round the edges to be estimated. A check may be obtained by changing the position of the grid and recounting the squares.

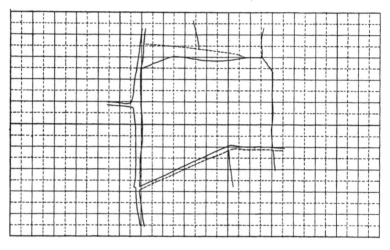

Fig. 6. A transparency ruled into acre and ¼-acre squares for computing areas on 6-inch maps by superposition. (Actual scale.)

Obtaining the figures from the 25-inch map has the great advantage that every area (even the stretches of road) has been accurately measured and it is therefore possible to make out a surface utilisation balance sheet for each parish. In this the total of the acreages for all the categories will agree with the total acres in the parish. The measured areas on the 25-inch maps will sometimes include two or more patches which need to be differentiated in the surface utilisation maps. In these cases the allocation of the acreage shown on the 25-inch map can be made with the aid of a planimeter or transparency. If the planimeter is used with care it will measure areas very accurately, but it is an expensive instrument and many people find it difficult to use with success. The transparency, on the other hand, costs nothing and gives quite as good results in untrained hands. Educationally it has the advantage that it develops the pupil's sense of area both on the map and in the field, a sense that is frequently vague in the extreme. The acreages of the various categories when they are calculated can conveniently be tabulated on the back of the map itself. A quarter sheet of the 6-inch map will usually include parts of two or more parishes. The acreages for each part parish should be tabulated separately. The tables for all the parts of a parish on different sheets of the map will, of course, give the table for the whole parish to be entered in the parish summary, which we shall discuss in the next chapter.

The making of the surface utilisation survey has a value both for the survey and for the surveyors. The need to look at every object shown on the map will reveal to the surveyor how much he did not know about the places he knew best, while he will acquire a fairly intimate acquaintance with an unknown district by this method in a surprisingly short time. A few years ago, in order to test the scheme now put forward, the authors set out, equipped only with the 6-inch maps, the 1-inch geological map and a note book, to survey together 24 square miles of country including a small town in a county unknown to them. They completed the surface utilisation survey in three fairly strenuous days and acquired in the process an intimate and detailed knowledge of the district that astonished themselves no less than the natives of the place.

Another advantage of the surface utilisation survey is the opportunity it affords for co-operative work upon definite, if elementary, lines at the very outset of a survey undertaking. It thus trains members of a survey organisation in the co-operative outlook which is so essential for carrying on a regional survey in its more advanced stages.

In addition to furnishing data for the statistical abstracts already mentioned, the surface utilisation surveyors will have collected data for the preparation of several of the maps mentioned in Chapter VIII. They will also have gathered a great deal of information that will be of service to workers in special branches of the survey. For instance, they will have mapped all the geological sections where the rocks are actually exposed. These the geologists will proceed to examine systematically. Again, they may have noticed suggestive banks or mounds in out of the way places which have hitherto escaped the notice of the archaeologists. The finished maps will be closely scrutinised by members working on any branch of the survey and notes made of places worth visiting from their various viewpoints. It is partly for this reason that observations not essential for the preliminary survey are always worth recording.

CHAPTER VI

THE INTENSIVE SURVEY OF PARISHES

THE method we shall describe in this chapter of carrying out the intensive survey of each of the parishes within the chosen area is a development of the surface utilisation survey. In those cases in which the survey area is of small extent and the base map is on the 6-inch scale, this method may advantageously be adopted for the surface utilisation survey itself.

The problem of making field notes and keeping them in a form in which they are readily available for reference is one that confronts all who are engaged in field researches. Such workers will evolve systems of their own which serve their purpose but their field notes are often unintelligible to others. Sometimes indeed the notes are unintelligible, after an interval, to the observers themselves, and the time spent in making voluminous but unorganised memoranda is found to have been wasted. In research work of any kind the difference between success and failure may often turn upon this question of method in making and recording observations. Particularly is this the case in regional survey work where the range of observations is so wide and team work so essential. The method here recommended, though it seems simple and obvious, has been evolved by the authors after some years of experience in the field and has been found to be very efficient in practice. It renders comparatively easy of accomplishment what appears at first sight to be an impossible amount of observation and record.

We shall deal with the survey of a rural parish to which the method is more particularly applicable. In the first place a tracing of the 6-inch map of the parish is made in drawing ink upon tracing cloth. This tracing should reproduce every *line* shown on the map. It will therefore show the hedges, roads, outlines of buildings, contours, etc., but not the "characteristics" and shading which indicate woodlands, etc. and private parklands. It is useful but not essential to mark upon it also the names of houses, farms, lanes, streams, woods, etc. which appear upon the original. For some districts geological maps on the 6-inch scale can be obtained from the Geological Survey, or referred to and copied in the library at Jermyn Street Museum. When the 6-inch geological maps are not available it will be necessary to copy the geological lines on to the tracing from the best available 1-inch maps (see Chapter IV). Different coloured waterproof inks may be used with advantage for streams, contours and geological lines.

Having completed the tracing *one* series of numbers is written upon it beginning at the left-hand top corner and giving a number to each area or object shown. For instance, each field, wood, pond, quarry, cottage, farm, house, inn and church will receive a number. If there is doubt whether an object shown on the map deserves a number to itself one should be given to it. An average parish will require from 400 to 500 numbers. The surface utilisation map, if it has been made, will be useful as a guide in numbering the tracing.

Instead of making a tracing the series of numbers may of course be written on the ordnance map itself, but the tracing has many advantages. The whole of a parish very rarely falls upon one 6-inch quarter sheet and often falls upon four or more. In these cases the tracing can be made upon a single piece of cloth. This will be strong, it will fold any way, spots of rain will not hurt it and on the whole it is well worth the initial trouble of preparation. If the base map is on the 6-inch scale it may be used instead of a tracing and several copies may be numbered for use by different surveyors. In any case a numbered copy should be kept, for otherwise if the tracing is worn out or lost the records will lose a great part of their value. If a tracing is made it will be a great advantage to have a few copies (with numbers) made by the true scale process mentioned in Chapter IV. Fig. 7 is a reproduction of a portion of the 6-inch map of the parish of Downe, Kent, with the numbers and geological outlines upon it. Owing to the difficulty of crowding the numbers into the small space occupied by the village itself on the map, an enlarged plan is made by the method described on p. 45, Chapter IV, or a tracing made from the 25-inch map. An enlargement of the village of Downe, four times linear, is shown in Fig. 8.

When the tracing or map is duly numbered the surveyor will take it into the field with a rough note book and commence to make his notes, writing the date at the head of them. Each note will consist of the number taken from the tracing followed by the observations, for instance:

281. Two small attached modern cottages, flint and brick, slate roofs, small gardens.

201. Pond for cattle, fed by road drainage, shaded on south by three ash trees, no pond vegetation.

113. Meadow, three horses grazing, much sorrel, mushrooms.

A road, hedgerow or ditch will be identified by hyphenating the numbers on either side of it:

179-200. Quickset hedge, maple and elm, 5 feet, toothwort abundant on elm-roots, one yew tree, one ash.

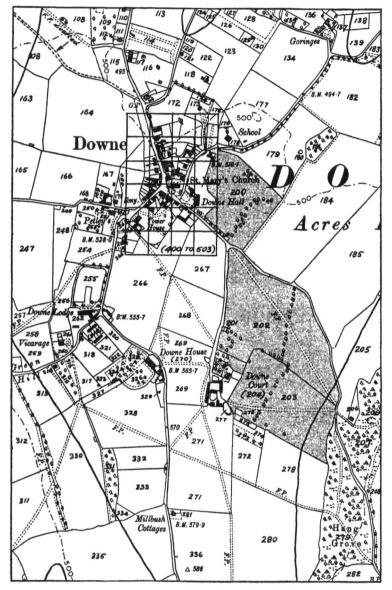

Fig. 7. A portion of a 6-inch Ordnance Survey map numbered ready for field work. (Actual scale.)

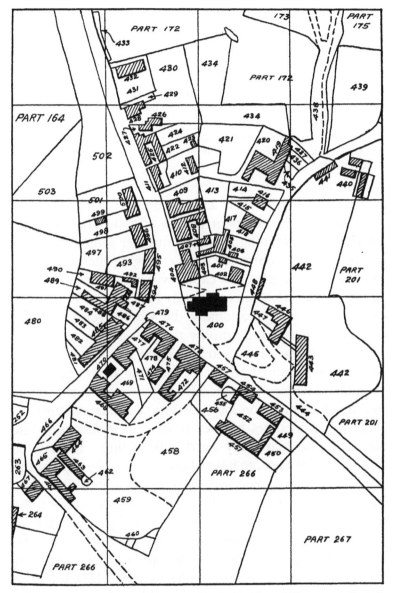

Fig. 8. An enlarged plan of Downe village (portion of Fig. 7, × 4 linear) numbered ready for field work.

The note made when observing a "number" for the first time should be not less than a sufficient description of the object or area to which it refers and should include suggestions if any for further investigation:

210. Strip of semi-natural chalk grassland, 1¾ acres, south-east aspect, fescue grass, orchids, gentian, thyme, eyebright, etc., suitable for ecological study and listing.

209. Small disused chalk quarry, section 7 feet, clean, layers of flint nodules, fossiliferous, floor of quarry overgrown with elder, bramble, belladonna, etc.

The notes made by different observers will naturally vary according to their knowledge and interests. They will, however, seek opportunities of going into the field together and learning from one another something of all aspects of the countryside, thus enlarging their spheres of observation, interest and capacity for co-operation in the work. In a school survey pupils may go out in pairs to make notes and occasionally the whole group with the teacher to check and amplify their observations.

In the course of the parish survey the names of fields should be obtained whenever possible from traditional and documentary sources. These frequently throw interesting sidelights upon local history (see Chapter VIII).

It will sometimes happen that what is shown as one area on the map may need subdividing. It may be found, for instance, to have changed since the date of the map, to contain some feature not indicated on the map or to include two or more crops. In most cases the necessary additions can be sketched in by eye or, failing this, the position on the map of any point in a field can be found by pacing. For this purpose it is not important to know the actual length of one's paces. The number of these *per inch on the map* can readily be computed by counting the number taken between any two points indicated upon it. It is useful to prepare a scale of one's paces for use with the 6-inch maps. The large field shown on the first specimen record card (Fig. 9), was divided by pacing into ten areas in less than a quarter of an hour.

When the surveyor returns home on each occasion with his notes, or as soon as convenient afterwards, he will transfer them from the rough field book into the parish record "book". This may either be a book or a series of loose index cards. Each form has its advantages but on the whole the authors favour the card index. If the book form is adopted the series of numbers should be written in the book allowing as much space for each number as appears necessary. The description of a church, for instance, will need much more space than that of a small cottage or a pond. If the card system is adopted this difficulty of estimating space will not arise. One or more cards will be used for each number. The cards will be taken into use as required and the space allotted to any number can be extended indefinitely by the addition of more cards. On the other hand, the lacunae in the observations are more readily detected in a record book. A loose-leaf note book affords a compromise but as already stated the balance of advantages is in our opinion with the cards in a vertical file. Photographic prints may be mounted on cards and inserted in their appropriate places as illustrations. A card index to the cards themselves showing under which numbers observations relating to the various sub-heads of the conspectus are recorded will greatly add to the value of the records.

Two specimen cards for the parish of Downe are reproduced in Fig. 9. Only the notes made upon the first visit are shown. Standard sizes of cards are $4'' \times 3''$, $5'' \times 4''$ and $8'' \times 5''$. The smallest size is not recommended for this purpose, especially when illustrations are to be included. The field notes will of course be

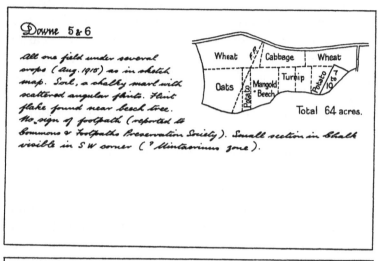

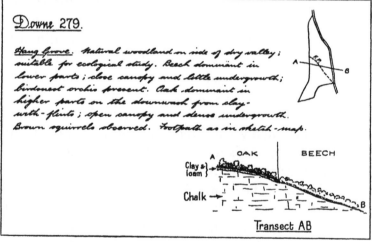

Fig. 9. Specimen record cards. (Originals 8″ × 5″.)

supplemented in many cases by extracts from, or references to, published matter dealing with objects in the parish, e.g. notes on the manor house from a county history or, on a geological exposure, from a geological survey memoir.

Readers who have attempted miscellaneous note taking in the field will readily appreciate the great amount of time and labour saved by adopting a method such as the one here outlined, which involves the minimum of paraphernalia. The tracing, a note book and pencil, are all that are needed. These can be taken in one's pocket whenever one is passing through the parish, even if the object of the excursion is not primarily for surveying. One great advantage of the method is that it is so well adapted to co-operative work. Any number of individuals, whether general observers or specialists in different branches of the survey, can contribute field notes to the same collection of parish records. In this way the record book or card cabinet will soon become a rich storehouse of information, valuable not only in itself but containing also the necessary data for the preparation of a large number of interesting maps of the parish. If the survey area contains several parishes these may either be allocated to separate workers or to subordinate survey groups, e.g. parish schools. Alternately one or two parishes may be done co-operatively before others are attempted. It is preferable to do one or two parishes thoroughly than to risk doing many only partially by too wide a diffusion of the available survey energies of the society.

When the parish survey is sufficiently advanced a summarised description of the parish can be written following the general plan of the conspectus. Fig. 10 is a reduced reproduction of the first page of such a summary of the parish of Downe, Kent. Whenever

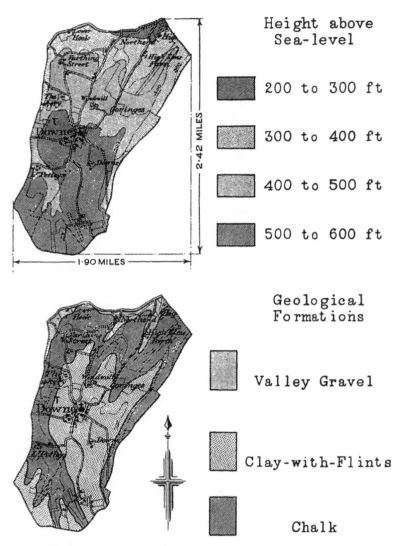

Fig. 10. A specimen of the first page of a parish summary.
(Coloured in original.)

possible, the data should be presented in simple maps and diagrams or in tables. It would be difficult by description to give a clear idea of the shape and relief of Downe parish, but the first little map accomplishes this admirably. Similarly, the second map, when supplemented by one or two geological transects, will furnish a better account of the geology of Downe than could be given in several pages of description. The base map of the Croydon Survey from which this instance is taken happens to be on the 1-inch scale, and it is very convenient to be able to cut up copies of it for the parish summaries. If this cannot be done it is very simple to make little maps of parishes from the 1-inch Ordnance Survey map.

On p. 76 we have given a statistical abstract of the surface utilisation in Downe parish for the year 1920. This will form the second page in the parish summary. When the acreages are reduced to percentages they furnish us with a basis for the comparison and classification of parishes.

Following the statistical abstract the summary will contain a concise account of the parish under each of the sub-headings of the conspectus. Long verbal descriptions should be excluded from the summary but it should include references to literature, etc., dealing with the parish or objects in it. It should also be liberally illustrated by miniature maps and transects and small photographic prints.

The method of surveying described in this chapter is specially adapted for predominantly rural parishes. It may be followed with modifications in town areas. Maps on a larger scale will be needed and the urban surface utilisation categories will need to be further differentiated.

The subject of the Civic Survey presents a whole set of new problems in technique, the treatment of which must be reserved for a future volume. It will be necessary there to deal with sections 700, 800 and 900 of the conspectus in much greater detail than is possible within the compass of this introduction.

Parish of Downe, Kent

Surface Utilisation, 1920. Statistical Abstract

Surface utilisation category	Area (acres)	Area (%)
Houses with small gardens	31	1·9
Houses with larger gardens	40	2·2
Public roads. Length 9¼ miles	25	1·5
Farm buildings and farmyards. No. of farms 7	3¾	0·2
Permanent grassland	854	51·7
Arable land	311	18·8
Market gardens, etc.	10	0·6
Small fruit	31	1·9
Orchards. 13 acres	(Included above)	
Heathland and rough pasture	Nil	
Scrub	4¼	0·3
Woodland	338	20·5
Water	1	} 0·2
Other categories	3	
Total	1652	99·8

Total acres in parish 1652·096.

CHAPTER VII

TRANSECTS AND RELIEF MODELS

THE relief of a region and its relation to the geological structure are of such fundamental importance in a regional survey that we shall devote the present chapter, in anticipation of the next, to the two methods of presentation indicated in the title. The preparation of the contour and geological maps from data already available in published maps are comparatively straightforward tasks with which we shall deal in the next chapter, but the methods of utilising these same data in the construction of geological transects and relief models call for some explanation.

We will consider first the simplest case of projecting the profile of a section across the contour map. The method will be almost self-evident from the example given in Fig. 11. The plan on the left shows the 50-foot contours only, extracted from the map of a portion of Surrey given in Fig. 21. It is desired to project the elevation of a section along the line XX. A base line $X'X'$ to represent sea-level is first set down parallel to XX and perpendiculars to it drawn at each extremity. Upon these lines we shall set off the *vertical scale* having first decided what height in the diagram is to represent 100 feet. We might make the vertical scale the same as the horizontal. If the horizontal (map) scale is 1 inch to 1 mile (5280 feet) then 100 feet will be represented on the vertical scale by about 1/53 of an inch. This, however, will not give us a very imposing profile and we shall do well to follow the usual practice of somewhat exagge-

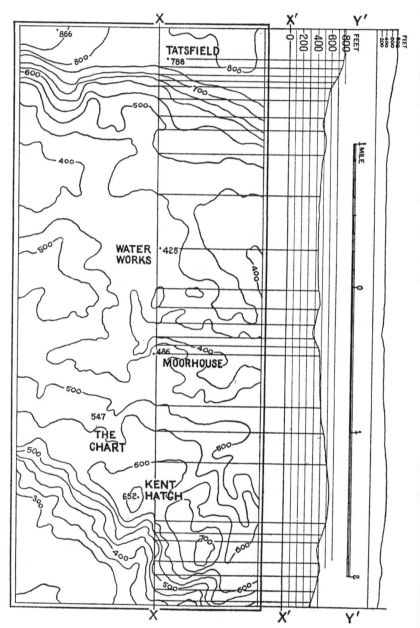

Fig. 11. Illustrating the method of projecting a relief transect.

rating the vertical scale. An exaggeration of the vertical scale by about two and a half times (i.e. 1/20 inch to 100 feet of height if the scale of the map is 1 inch to 1 mile) is very appropriate, and with a little practice it will not be found difficult to construct good sections on this scale. We have adopted it in the sections in Figs. 11 and 12, and in the former we have inserted the true scale profile below for comparison. Although the true scale is a faithful reproduction of nature it does not *look* so true as the exaggerated profile. It is perhaps not superfluous here to emphasise the need for using a finely pointed hard pencil (3H) in the construction of transect diagrams.

Having set off the vertical scale on the perpendiculars $X'Y'$ we may rule faint pencil lines parallel to the base $X'X'$ at each 100-foot interval. Then at each point where the line XX intersects a contour line on the map we shall raise a perpendicular to meet the base line $X'X'$ and extend it to the pencil line representing the corresponding level. When this has been done for every point of intersection we can sketch in the profile curve by joining up the tops of all the perpendiculars. The more perpendiculars we have the more perfect will be the curves of the profile. Thus, if the map has contours at 50-foot intervals, as in our example, we shall be able to draw twice as many perpendiculars from the line of section and so determine twice as many points on the profile curve. This is one reason for preparing a 50-foot contour map, as directed in the next chapter, if the base map itself does not show 50-foot contours.

In addition to the contours the 1-inch and 6-inch ordnance maps give the heights in feet above sea-level at several points along roads and elsewhere. If any of these points are on or near the line of section they will give some indication of the course of the profile curve between the perpendiculars. Four such additional points occur near the line of our section, namely, Tatsfield 788, Water Works 425, Moorhouse 486, and The Chart 547. An examination of the trend of contour lines on either side of the line of section will also help us to sketch in the profile curve correctly.

We may now proceed to the more complex problem of constructing a geological transect. Its difficulties are not insurmountable even by those who are not trained in geology, and the making of these transects is so important, both in the study of the region and in expressing the results of such study, that the methods of their construction should be mastered by all regional surveyors whether or not geology is their special interest.

The plan on the left in Fig. 12 shows, in addition to the contours in Fig. 11, the geological lines extracted from the map on p. 114. We have already constructed a profile section, and now wish to add to this the internal geological structure. A geological map represents the rocks composing the earth's crust as they appear at the surface, or rather as they would appear if we could remove the covering of soil. The area occupied by any one geological stratum or formation at the surface is known as its outcrop and the lines on a geological map are the boundaries between adjacent outcrops. In our example we have the following outcrops commencing from the bottom of the map, namely, Weald Clay (h^1), a narrow outcrop of Atherfield Clay ($h^{2'}$), a broad stretch of Hythe Beds ($h^{2''}$), then Folkestone Beds ($h^{2'''}$) of which a small "outlier" has been left upon the Hythe Beds at Moorhouse, next Gault (h^3), Upper Greensand (h^4) and finally Chalk (h^5) at the top. The heavy broken line re-

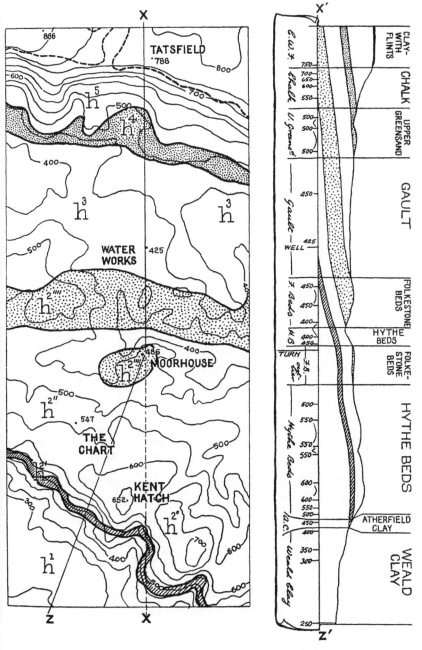

Fig. 12. Illustrating the method of projecting a geological transect.

presents the boundaries of a superficial deposit of Clay-with-flints (C.W.F.) which rests upon parts of the chalk area. On a geological map the different outcrops are usually indicated by different colours or shadings to aid visualisation, but we have here inserted only the lines with some stippling and the letters used by the Geological Survey to denote the various formations. The several outcrops bearing the designation "h^2" together constitute the Lower Greensand formation while all the formations in our example bear the letter "h" which indicates that they all belong to what is called the Cretaceous System.

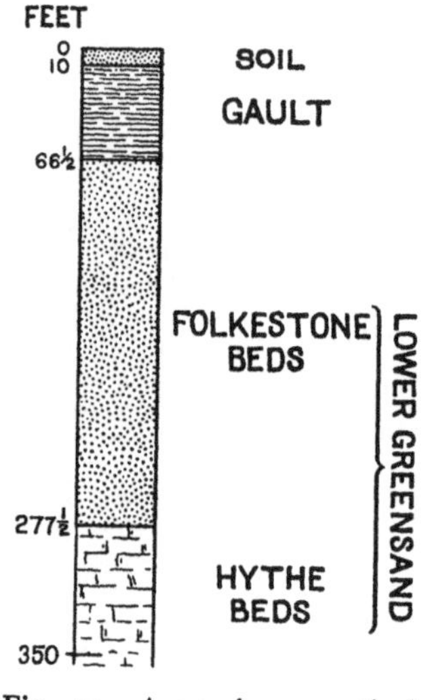

Geologists distinguish two kinds of *sections* known respectively as vertical and horizontal. Vertical sections indicate merely the order and thickness of the strata below a given point. The strata passed through in making a well-boring or mine shaft, for instance, may be represented diagrammatically in a vertical section. Thus Fig. 13 is a vertical section indicating the strata found in boring the well at the waterworks, near Titsey Wood on the map, as recorded in the Geological Survey Memoir on the Water Supply of Surrey.

Fig. 13. A specimen vertical geological section. (The well section, Figs. 12 and 14.)

A horizontal section or *transect*, as it is now more appropriately termed, represents the geological strata as they would appear if the crust of the earth could be cut vertically along a given line and the cut surface exposed to view as in the block diagram on p. 99. Nature sometimes furnishes us with good transects in sea cliffs, as for instance at Alum Bay in the Isle of Wight, but it is less easy to insert the geological strata accurately below the profile curve of an inland section. The first step, namely, marking the positions of the outcrops at the surface is quite simple. We have only to raise perpendiculars from the points where the line of the section crosses the lines of outcrop on the map, as already explained in the case of contours. It will be noted, however, that in Fig. 12 we have bent the line of section near the middle so as to alter its direction. It is often expedient to do this in a geological transect in order to produce a characteristic section. In the present case it will be seen that the section along the line XX in Fig. 11 fails to depict the escarpment of the Lower Greensand in the clear way in which it is brought out by the slight bending of the line of section as in Fig. 12.

But although the line of section may change its direction or even pursue a zig-zag course on the map in order to bring out the required features, or to include or exclude some local peculiarity, the transect diagram must necessarily lie in one plane, that of the drawing paper.

We have thought it expedient to explain the geometrical method of constructing a profile by means of Fig. 11, but in practice it is not usually done in this way. The base line $X'X'$ is drawn on a separate sheet of paper and the points transferred to it from the line of section on the map by dividers or by marking them off on the straight edge of a piece of paper. Thus in Fig. 12, instead of projecting geometrically the points on the line of section, we have depicted a strip of paper with these points marked on its edge applied to the base line $X'Z'$.

This method eliminates the complication of projecting points from a bent line of section and will enable us if necessary to take the contour heights from one map and the geological outcrops from another. In this case the line of section XX must of course be carefully drawn in the same position on both maps.

It still remains for us to indicate in the transect diagram the disposition of the strata beneath the surface. Very occasionally they will be disposed horizontally but more often they are tilted, the angle of tilt being termed the *dip*. But how can we know from an inspection of the surface whether they are level or tilted and if tilted at what angle? We should here be confronted by a very difficult problem if we had to rely upon direct first-hand observation. A trained geologist can often make a shrewd guess at the internal structure from the conformation of the surface and he could confirm or modify his conclusions by an inspection of the rocks exposed in quarries, or railway or road cuttings he could find in the neighbourhood. But let us see what we can do even though we are not trained geologists.

If we were fortunate enough to have records of well-borings at, say, ½-mile intervals along the whole line of our section we could plot out the underground positions of the strata as we have done for the waterworks (see WW', Fig. 14 *a* and *b*). In the absence of such data we can only fall back upon conjectures based upon our general knowledge of the geology of the neighbourhood. From the data disclosed in boring the well at the waterworks and the known position of the outcrops near by of the strata in the well section, we can draw in the underground structure between the waterworks and the outcrops. Thus the lines $A'A$ and $B'B$ in Fig. 14 *a*, give us the underground position of the Folkestone Beds between the Gault above and the Hythe Beds below. They also indicate the dip of the strata. If only we could rely upon this dip being uniform in the neighbourhood it would be a very simple matter to complete our transect, as we have done in Fig. 14 *a*, by producing the lines AA' and BB' and drawing others parallel to them from the other geological points C, D, E and F. But we never

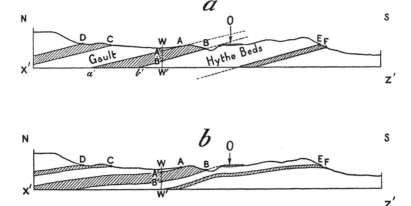

Fig. 14. Illustrating the insertion of the strata in a geological transect.

can depend upon the dip being constant and the transect (*a*) constructed upon this assumption is far from being an accurate representation of the strata beneath the line of section. They are in fact much more nearly as shown in the transect (*b*). Not only does (*a*) give a greatly exaggerated thickness for all the beds but it will be seen that if the line $B'B$, i.e. the base of the Folkestone Beds, is produced southwards, it misses entirely the Folkestone Bed outlier O. This indicates that there must be a bend in the strata between B' and O and we happen to know from other considerations that this is so (see transect (b)). This example is a comparatively simple case. The transect will sometimes be complicated by underground contortions of the strata or dislocations of them by "faulting". The making of geological transects, which we regard as so essential in the study of a region, would often be a very difficult matter but for the fact that geologists have already made a careful study of the structure of the British Isles and have published not only maps but transects for every locality. There is at least one horizontal section on many of the sheets of the Geological Survey map[1] and several others published on separate sheets. Others will be found in Geological Survey Memoirs and many more in various geological textbooks and other publications such as the *Journal of the Geological Society* and the *Proceedings of the Geologists' Association*. A regional survey organisation should acquire or make copies of all the published sections which throw light upon the geological structure of its region, and when these have been carefully studied in conjunction with the geological map the construction of original transects will be found to be a very instructive and not too difficult task.

In the case of a school survey when a profile has been constructed by the pupils as an exercise in practical geometry, and the geological strata have been inserted with such help as the teacher may give them in a lesson in general science, they may utilise the occasion of a "school journey" or other excursion to follow out the line of the transect in the field. On such an occasion the pupils will examine exposures of the rocks in quarries and road cuttings and collect samples for future examination, perhaps as exercises in chemistry and physics. They will note

[1] See Fig. 3 a, p. 41.

such things as the kinds of trees, the crops and soils to be found at different parts of the transect; the points where springs emerge, the directions and manner in which the streams traverse the various outcrops and gather every kind of information that will enable them to build up a diagram like the example we have given in Fig. 22, p. 115.

We have seen how the geological transect enables us to depict what may be called the anatomy of the region. By adding other data to the transect, as in Fig. 22, we are able to express something of its physiology. The block diagram on p. 99, gives us an idea of its morphology so far as this may be portrayed in a plane figure. A block diagram may be regarded as a perspective drawing of a relief model. It will enable us to visualise the topography and structure of the region better almost than the model itself, but a relief model has many other merits which make its construction very desirable at an early stage in the progress of a survey. A carefully constructed model will present to us in a single view the relief of the whole region in its broad features and in its detail. It will be of great value for purposes of both demonstration and study.

There are various methods of making relief models. They may be carved out of wood or any other suitable material, modelled in a plastic medium or built up in layers. The first method calls for considerable skill in carving. The second is less difficult and may fittingly be adopted as a school exercise. The body of the model may in these cases be built up of layers of thick card or linoleum and the "steps" filled in with modelling clay. Both these methods, however, have the great disadvantage that the relief will need to be greatly exaggerated in order to give the desired effect in the finished model.

We recommend very strongly the adoption of the third method which we shall describe, namely, building up the model in layers. Models made in this way will give a very striking effect of relief *even if the vertical scale is equal to the horizontal.* In practice it may be found convenient to increase the vertical scale to about one and a half times or twice that of the horizontal. Minute accuracy is much less dependent upon skill in the case of layered models than in carving or modelling. Professor Hawkins of Reading University has raised the construction of layered models to a fine art. He has constructed some very beautiful models of wide areas on the scale, horizontal and vertical, of 1 inch to 1 mile. In order to achieve these results he has taken an amount of pains which probably few will care to emulate. In the first place he has, from field observations, drawn in the contours on the map at intervals of 10 feet. Each layer in one of his models represents 10 feet altitude, so that a model of the Snowdon district will contain some 350 layers which mount up to only two-thirds of an inch in thickness. The layers are of thin cartridge paper carefully selected with the aid of a micrometer gauge. He has a method of soaking each shaped piece of paper in thin paste before placing it in position so that it adheres when dry. As the wet paper is slightly expanded in area, great care is taken to so place each layer that it will shrink into its correct position upon drying. Professor Hawkins' models are objects of surpassing beauty but the amount of patience and skill bestowed upon them seems almost incredible to the ordinary mortal. We may remember, however, that *le mieux c'est l'ennemi du bien* and we need not despair, for very useful and pleasing layered models can be made with far less expenditure of time and care. We have made such models in a week of evenings and found the process very fascinating. Given a contour map they can easily be made by school children, the layers being cut out of thin card of about 1/60 inch in thickness such as is used in schools for various purposes.

With few exceptions, published maps do not show contour lines at more frequent intervals than 50 feet. These are shown on the popular edition of the Ordnance Survey map (see Chapter IV, p. 37). If the area under survey is considerable a model on the 1-inch scale with layers, each representing 50 feet of altitude, will be well worth the making. The result will probably arouse sufficient enthusiasm to determine the model maker to obtain intermediate contours by field observation in order to make a still better one. The closer the contours and the greater the number of layers the more perfect, of course, will be the finished model. It is comparatively easy to insert 25-foot contours between the 50-foot lines already shown on the map. We may call these "form lines" seeing that they will not be so mathematically accurate as the contours inserted by the use of "levelling" instruments. A model on the 2-inch scale with layers representing 25 feet will be a great improvement upon the above-mentioned 1-inch model. By using an aneroid, or other form of altimeter, form lines at $12\frac{1}{2}$-foot intervals can be drawn in on the map. For this purpose it will be desirable to use the 6-inch maps upon which the 50-foot contours have been inserted in manuscript. A model on the horizontal scale of 3 inches to 1 mile with layers representing $12\frac{1}{2}$ feet will give an excellent result. The thin card already mentioned is suitable for building either of these three models and in each case the vertical scale will be found to be about twice the horizontal scale. Fig. 15 is a

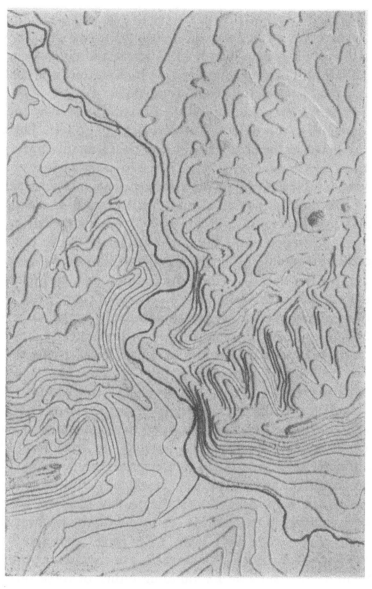

Fig. 15. A layered relief model of the Mole Gap, Surrey.
(Scale 1 inch = 1 mile; layers at 50-foot intervals.)

facsimile of a layered model of the Mole Gap on the 1-inch scale with 50-foot layers.

The following instructions should be followed in constructing one of these models. A piece of thick ply-wood will form a suitable base for the model. Even this will tend to warp with the tension of the drying cards, and it is advisable to reinforce it by screwing it to a strong board or battens until the finished model is perfectly dry. The first sheet of card to be affixed to the base will be rectangular and will entirely cover it. It will represent the lowest altitude in the survey area. In order to get the other layers it will first be necessary to make a tracing of the contours from the map. Care should be taken to select a piece of tracing paper that will not expand and contract according to the amount of moisture in the atmosphere. From the tracing the contours may be successively transferred to the cards by the use of carbon paper and a sharply pointed hard pencil. It is, of course, only necessary to trace the one contour in question on to the card for each layer, but care must be taken not to miss that contour in any part of the tracing. It is sufficient to trace a few small portions of the next higher contour as a guide in placing the next layer in position on the model. Thin paste is the most suitable adhesive. When the model is finished and quite dry its surface should be brushed over with size and a coat of flat white paint applied. The surface of the model will be "terraced" and it is just this feature that produces the desired effect of relief. Nothing should be drawn on the surface of the model or this effect will be marred.

The edges of the complete model may be rubbed down with glasspaper, or made perfectly true by skilful planing. They should then be painted white, after which the geological strata may be depicted upon them as indicated in the block diagram, Fig. 17.

The scale of the model will of course be determined by the magnitude of the area. The model can be made any size but for several reasons it is undesirable that it should exceed about 15 inches square. There is the difficulty of making large sheets of stiff card adhere satisfactorily without proper pressing devices and the tendency to warp increases enormously with increase of area. On the 3-inch scale, a 15-inch square would represent 25 square miles, on the 2-inch scale, 56 square miles and on the 1-inch scale, 225 square miles. If the desired model will materially exceed the above dimensions we should advise its being made in two or more sections.

CHAPTER VIII

THE REGIONAL SURVEY ATLAS

"In Geography we may take it as an axiom that what cannot be mapped cannot be described."
   DR HUGH ROBERT MILL, Herbertson Memorial Lecture, 1921.

WE have already made it clear that the basis of the accumulating records of a regional survey will be an ever growing collection of maps depicting the results of field work in the various branches. These maps, numbered and arranged as indicated in the extended conspectus (pp. 18 to 20), may be conveniently kept in a loose-leaf album so as to form an atlas of the region. In the quotation which heads this chapter we should not be quite justified in substituting "regional survey" for "geography", but we cannot too strongly emphasise the fact that whenever possible the most satisfactory method of recording regional observations is on a map. Nearly always it is possible but the beginner should guard against the mistake of trying to show too much on a single map. The atlas will not, however, be composed entirely of maps. It will contain also, on sheets of uniform size, transect and block diagrams (see Chapter VII), statistical tables, enlarged plans of village sites, earthworks, etc., and brief supplementary descriptions of the maps when the marginal space of the map sheet is insufficient for this purpose. Longer written theses will be more conveniently filed separately.

The number of maps that may be prepared is almost unlimited and we must of necessity confine our attention chiefly to those which we consider essential in any survey and some others which may be prepared by workers having no technical knowledge of special subjects. If a start is made with these many others will suggest themselves as the work proceeds. At the close of this chapter we have set out in the order of the conspectus a longer list of suggested maps and diagrams.

In making the following suggestions we shall depart from the order of the conspectus. That order is determined by the logical considerations summarised in the diagram on p. 10, but the actual survey work and the preparation of maps will not in practice follow it. Nor will the regional phenomena present themselves to the surveyor, still less to the schoolboy, in that sequence. Our everyday observations are not guided by considerations of cause and effect, the discovery of which is the province of science. For instance, although the geological structure of a region is the primary physical determinant of all its features it will not thrust itself upon the eye of the observer. He will notice first such things as the relief, the location of urban and rural areas, the rivers, roads and railways and it will not be at all self-evident that these are all closely related to the geological structure, our scientific appreciation of which is scarcely a century old.

For our purpose we can conveniently divide the maps that may be prepared into three groups, namely:

(*a*) Maps that may be made without leaving the study from data already available (e.g. relief, population).

(*b*) Maps in which the compilation from available data will need to be supplemented to a greater or less extent by field observations, but which will not require special technical knowledge (e.g. woodlands, public open spaces, industries).

Fortunately a very large number of maps fall within these two groups.

(c) Maps for the preparation of which technical knowledge is necessary (e.g. types of natural vegetation, soils, earthworks).

In most subjects there will be maps in each of these groups. For instance, a preliminary map of the woodlands may be made by simply colouring the woods indicated on the base map. It will, however, be necessary to examine the woods themselves before they can be differentiated into oak, beech, pine, larch woods, etc., on the map, though this will scarcely need the help of a botanist, as will the more advanced ecological study of the woodlands. A preliminary "group a" map is always desirable, even though it may have to be cast aside when field observations have been completed. In the case of some of the leading maps it is useful to make additional copies on transparent tracing paper. These will be employed for placing over other maps for comparison. The distribution of population, for instance, with regard to relief, or of woodlands or crops with regard to geological outcrops, may readily be tested in this way often with illuminating results.

We shall assume throughout that the survey area is a fairly extensive one including several civil parishes and that the base map is on one of the smaller scales. Most of the suggested maps are, however, equally appropriate for small survey areas of one or two parishes with larger scale base maps.

Quite a number of maps can be prepared straight away by the method known as map-analysis. This consists in selecting one of the many topographical features—contours, roads, rivers, civil parishes, woods and so on—and either preparing a map to show that particular feature alone or, by the use of colour, bringing it into visual prominence on the base map at the expense of the other features.

Relief. The relief of the region may well form the first line of approach in a new survey, partly on account of its primary importance in the study of a region and partly because the data are to hand in the contour lines shown on published maps. In a school survey every pupil should be encouraged to master the relief of the district by exercises in making contour maps, transect diagrams (at first without geological structure), and relief models. Whichever of the various map issues discussed in Chapter IV is chosen for the base map, the contours at 100-foot or 50-foot intervals will appear upon it. A relief map can therefore easily be made by accentuating these lines with drawing ink and colouring the spaces ("layers") between them. The most suitable colours for this purpose are green for the lowest levels decreasing in intensity in the ascending layers, merging into yellow for the middle heights and brownish yellow to dark brown for the highest altitudes. The sea and expanses of inland water will be coloured pale blue. These colours may be produced in graduated shades by diluting and mixing Winsor and Newton's "Regional Survey Colours". The suggested colour scheme approximates to conventional cartographic usage but gives more emphasis to the relief than do the less vigorous tints of printed maps. If the base map shows only the 100-foot lines, the intervening 50-foot lines should be inserted on one map for the survey collection. The contour map may be usefully supplemented by a table giving the areas of the several "layers". It is also one of those of which a copy should be made upon transparent paper.

If the base map is on a small scale it will be found very

profitable to prepare a large relief map on the scale of, say, 3 inches to 1 mile, the contour lines being compiled by reduction from the 6-inch maps. Such a map is invaluable for purposes of demonstration and it will exhibit a degree of accuracy and detail which is not attained in small maps. It will also prove very useful in the preparation of large scale transect diagrams for teaching or exhibition purposes.

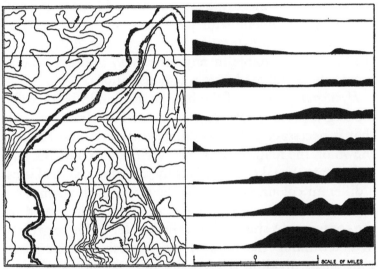

Fig. 16. Silhouette transects (reduced), Medway Valley.

Relief transects in the form of black silhouettes, omitting the geological strata, are instructive. A series of such transects at ½-mile intervals across the map or any part of it, e.g. a valley (see Fig. 16), will help the student to grasp the relief and will sometimes disclose interesting features that would otherwise escape notice. Silhouette transects of the main roads have a special appeal to cyclists and motorists and form useful supplements to the "Communications" map.

The preparation of a relief model, as described in the preceding chapter, should be undertaken at an early stage of the survey. Even a rough model will be found to be a tremendous asset to surveyors in every branch of the work. Those who, after long procrastination and doubt, have discovered how simple the process is, have always regretted that they did not make the attempt sooner.

The drawing of "block diagrams", as illustrated in Fig. 17, is a little more difficult but it should not be beyond the power of anyone with some aptitude for sketching.[1] This mode of representation gives us a view of the region that is not attainable in any other way and one that is especially useful for lantern demonstrations.

Hydrography. One of the base maps should be used for making a preliminary map of the surface drainage. All that is necessary is to accentuate the rivers and streams shown on the map by drawing over them in blue ink and to indicate the lakes and ponds by the same colour. Using a different colour, say brown, the water partings may be inserted. These will pass along the ridges as indicated by the contours. A study of the relief model will help greatly in determining the lines of the watersheds. The main watersheds dividing river basins may be indicated by a continuous line and the minor partings separating the catchment areas of tributary streams by broken lines. Some of the streams will start on the higher ground as surface drainage while others will have their sources in springs. The positions of springs are marked on the 6-inch maps from which they should be copied on the drainage map. When the surface utilisation survey has been made we shall be able to revise the drainage map,

---

[1] A very useful treatise entitled *Block Diagrams* has been written by A. K. Lobeck.

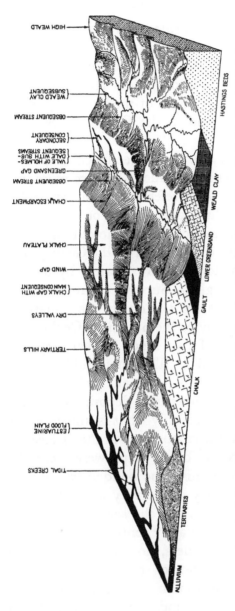

Fig. 17. A block diagram illustrating the structure of North Kent. (From Ogilvie's *Great Britain*.)

if necessary, and to indicate upon it by blue stippling the marshy or boggy areas. Another map can be used to mark the positions of all the wells in the region, differentiating shallow from deep, and recording the level of the water in them. Much of the necessary information for this will be found in the Geological Survey Water Supply Memoirs.

Geology. In every survey the geological maps are among the first that should be made. They are more important and will be more often referred to than any other map. If the geological lines are already on the base maps it will only be necessary to colour the outcrops, following, for preference, the colours used by the Geological Survey. These can be matched approximately by using Winsor and Newton's "Regional Survey Colours" (see p. 55). If the geological lines are not on the base maps no time should be lost in copying them from the Geological Survey maps. These are of two kinds known respectively as "Solid" and "Drift". The first published series of Geological Survey maps were all "solid", that is they took no account of superficial deposits of gravel, brickearth, etc., but showed only what is known as the solid geology. As this is often obscured at the surface by recent "drift" deposits the older maps were somewhat misleading to agriculturalists and builders who were advised to consult them. They served very well the needs of civil and mining engineers, and the scientific interests of the older geologists were mainly centred in the solid geology. The farmers and builders, however, not unreasonably, complained when a district shown as chalk on the map turned out upon inspection to be gravel or clay and the Geological Survey was induced to issue a second series of "drift" maps showing the superficial deposits where these occurred over the older formations. In more recent times the attention of geologists has been increasingly directed towards the drift deposits, partly because of the light they shed upon the development of our river systems and partly because of the stone implements used by early man which are often found embedded in them.

To the regional surveyor all these aspects of geology are of importance and it will therefore be necessary to prepare and colour both a solid and a drift map. The solid map will give the key to the general structure of the region and furnish the data for making transect diagrams, while the drift map is the essential basis for the study of such aspects as types of vegetation, soils, agriculture, prehistoric man, etc. As in the case of the contour map a tracing of the drift map should be made on transparent paper.

If the base map has not the geological lines they should be drawn in from the drift map on a number of copies to be used for various purposes. Several maps, e.g. vegetation, rivers, mineral industries, will gain much in significance if the geological lines are present also.

An uncoloured copy of the geological map should be used to mark the positions of all geological exposures (quarries, road and railway cuttings). This can be done more satisfactorily when the surface utilisation survey has been made. A number should be given to each exposure on the map and a description from field observations written on a separate sheet or index card. For the sake of uniformity it is convenient to use a form for this purpose. A specimen of the form used in the Croydon Survey is here reproduced (Fig. 18). It will serve to indicate the kind of records that are required. It frequently happens that geological exposures are of a very

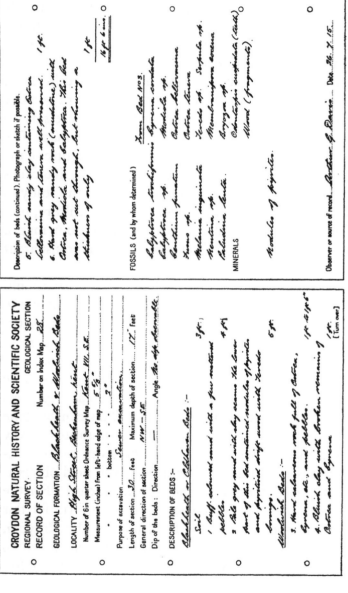

Fig. 18. A specimen record of a geological exposure. (Original, small 4to size.)

temporary nature (e.g. excavations for foundations or for drains). Every endeavour should be made to record particulars of these as they occur. Records of exposures no longer visible will sometimes be available in various publications. These should be extracted on to forms and added to the survey collection.

Another map, or the same one using a different coloured ink, should be used as an index to well sections and mine shafts, the particulars of which, preferably in the form of vertical section diagrams may be recorded on a separate sheet or sheets. These particulars can usually be obtained from Geological Survey Memoirs or if unrecorded from the engineer at the waterworks or factory where they occur.

Still another geological map should be used to show and index the lines of transects which have either been prepared especially for the survey or copied from Geological Survey or other publications. The transects themselves, numbered to correspond with the lines on the index map, should be drawn on sheets of uniform size with the map sheets and inserted in the atlas. The more of these that can be made or collected the better. A sheet of transects for a series of parallel lines drawn at $\frac{1}{2}$-mile interval across the base map roughly at right-angles to the geological strike will be found to be very illuminating.

Meteorology. For recording atmospheric phenomena (rainfall, pressure, temperature, wind) some simple apparatus is necessary. Full instructions for making and using such apparatus is given in an admirable school science manual entitled, *The Study of the Weather*, by E. H. Chapman. The preparation of meteorological maps based upon original field observations involves the setting up of a number of stations each with appropriate apparatus at which daily readings can be taken, but it is possible to prepare some useful preliminary maps from existing records such as are issued from the Meteorological Office, now controlled by the Air Ministry. Maps showing the average annual rainfall (isohyets at $2\frac{1}{2}$-inch intervals) have been published for many parts of the British Isles (see the Royal Meteorological Society's *Rainfall Atlas of the British Isles*, 1926, and the Geological Survey Water Supply Memoirs for various counties). These data should be copied on one of the base maps.

Fig. 19 is an interesting example of what can be done with the help of school children. It is a map of the course of a violent hailstorm which occurred in Surrey in 1918, and was prepared by Mr J. E. Clark, largely from data (damage done) collected from children by schoolmasters in the district. It is here reproduced by the courtesy of Mr Clark and of the Royal Meteorological Society.

Vegetation. We have already mentioned that a preliminary woodland map may be prepared by colouring the woods shown on the base map. Beyond this little can be done without field work except in the few districts for which vegetation maps have already been published. When the surface utilisation survey has been made the woodland map may be improved by showing the types of woodland (indicated by the dominant trees), and some other useful features (e.g. heaths) added to it. The regional surveyor's interest in vegetation extends far beyond the compilation of a flora of the region, that is its content of plant species. It embraces also the study of the social communities of plants, their composition, development, distribution and relation to the many factors which constitute their environment. These aspects of vegetation comprise what is now called "Plant Ecology". The preparation of a good vegetation map which is much to be desired, will necessitate extensive field study by trained ecologists. Surveyors whose interest lies in this direction are strongly advised to study *Practical Plant Ecology*, by Professor A. G. Tansley and *Types of British Vegetation*, edited by the same author. The latter is out of print,

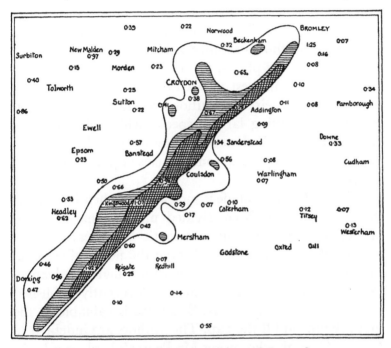

Fig. 19. Map of Surrey hailstorm, July, 1918.

but may be seen in many libraries; the former, a more recent work, contains a very useful bibliography and list of published vegetation maps.

An uncoloured geological map can be utilised to show the areas where natural or semi-natural vegetation still exists in the region, and for as many as possible of these full lists of the species occurring in them with their relative abundance or scarcity should be compiled. This is quite straightforward work that is within the power of any botanical student. Not only will these lists form a necessary foundation for a more advanced ecological survey but they will prepare the makers of them for a full appreciation of the ecological method of studying vegetation. It will be seen from them that the groupings of species on like soils are similar to one another and different from those on other soils.

Some species of plants, for one reason or another, will be of sufficient interest to warrant separate maps being used to show their distribution (for instance, the food plants of interesting larvae: species which play the part of hosts to fungi which at other times in their life cycle are destructive of crops: and some species of purely botanical interest).

Animal Life. Animal life is such a comprehensive subject that its study has been divided up into many special branches. Moreover, the systematic ecological study of animal life, which alone will serve the needs of regional survey, has not yet reached the same stage of advancement as in the case of vegetation. A recently published book, *Animal Ecology*, by A. S. Pearce is valuable as the first attempt to fill the gap. The authors are indebted to Miss E. C. Pugh who, using Mr Pearce's book as a basis, has worked out section 600 of the conspectus. In *Animal Ecology*, the author has discussed the general principles of the subject, but in any given region the application of these principles to the study of the animal life will still be found to offer virgin soil to pioneer workers.

A start may be made by workers who are not specialists by mapping the distributions of some of the more conspicuous mammals and birds, such as rabbits or rooks, but beyond this the help of field naturalists will be needed for this branch of the survey.

History and Archaeology. The roots of the future are in the past and the life of a region as we see it at any given time presents a mosaic of survivals and developments from the past together with incipient tendencies foreshadowing the future. Roman roads and Roman law still survive in our land, our churchyards are often British, our churches and much of our village life are Mediaeval, our domestic architecture reflects the culture of all ages from Mediaeval down to our own, our agriculture is Neolithic, Saxon, Georgian and again Georgian, our surnames come down to us from the twelfth and thirteenth centuries, our language testifies to successive invasions and cultural contacts—British, Roman, Saxon, Norman and in our day American, and so we might go on.

The difference between history and archaeology, it might be said, depends upon the period dealt with, but it is still more a difference of method. Broadly speaking, the archaeologist gathers his material from investigations in the field while the historian pursues his researches chiefly amongst documents. The period before to the coming of the Romans in this country is the domain almost exclusively of the archaeologist, but from that time onwards the field is increasingly open to the historian also, and the two methods of investigation should supplement and check one another in the search for truth. The historian has the Itineraries of Caesar, but the archaeologist alone can essay to mark them down with geographical precision; the historian knows from documentary evidence that a mediaeval manor house once stood in a certain parish but its site may perhaps only be definitely known when the archaeologist by excavation has discovered its buried foundations.

The regional historian will pursue his investigations by both methods. He will not neglect to study the usual documentary sources of historical information such as are enumerated in Rev. J. Charles Cox's well known and admirable little book, *How to Write the History of a Parish*. But he will also follow the example of the archaeologist and search in the region itself for historical evidences.

The natural course of a regional survey is such that it will tend to write history backwards beginning with the known, or at least observable present and pressing back from it into the less known past. Let us take as a simple illustration the question of population. We should naturally choose the latest census publication for the purpose of preparing our first population maps. We shall then go back to the preceding census and so on by ten-year steps right through the nineteenth century. Again let us take the allied subject of housing. Given a large-scale map of a town area we should prepare a historical housing map by first colouring the houses built since the close of the war, next those erected between 1900 and 1914, following with older buildings in ten- or twenty-five-year intervals until by a process of elimination we have exposed to view the survivals from earlier times.

Economics and Sociology. Coming now to the human side of the survey a few essential maps can be made by simple map-analysis. Of these the road, railway and civil parish maps are the most obvious. It is only necessary to accentuate each of these by colouring on a separate map, though ways of elaborating and improving the maps will no doubt be devised. For instance, a simple but effective map showing accessibility to railways may be made by describing circles at, say, ½-mile intervals with the railway stations as centres until the circles for adjacent stations meet. Thus a series of zones of increasing remoteness from railway stations will be produced. These zones may be coloured with washes of green in increasing density. Similarly a map showing accessibility to omnibus routes may be prepared by drawing lines parallel to the routes and in a third map these two may be combined. Such maps will be of especial significance in relation to the distribution of population, the development of building sites and other social and economic features of the region.

With the help of the Ordnance Survey County Administrative Map or Kelly's *County Directory*, a map showing by differential colouring the municipal or county boroughs, urban and rural districts can be prepared.

From the census publications (which are published separately for each county) population density maps can easily be compiled. The census publication will give the populations of whole parishes only or in town areas of wards. If these totals are divided by the number of acres in the parish or ward the relative densities of population thus disclosed can be indicated on the map by suitable washes of colour. In the 1921 census reports the densities are already worked out for each parish.

By studying the distribution of houses on the 6-inch maps, or still better the surface utilisation maps, it will be possible to distribute the population approximately within the parishes and thus prepare on one of the base maps a much more accurate population map. The best method of representation for this purpose will be what is known as a "dot-distribution diagram". A dot will be placed on the map for each 10, 50 or 100 people as may be decided by the order of figures to be dealt with. Fig. 20 is a specimen dot-distribution diagram reproduced from *Great Britain* by the courtesy of the publishers.

The preparation of a few of the maps above described at an early stage will stimulate interest in the survey and reveal its possibilities to those whose imagination does not otherwise encompass them. A good maxim in regional survey is to do what you can without waiting for perfection. The first attempts can always be discarded when there is something better ready to take their place.

When the surface utilisation survey has been completed data will be available for quite a number of special maps. One of the base maps can now be made into a generalised surface utilisation map with fewer categories than on the 6-inch maps. Such a map differentiating only urban areas, agricultural land, woodland, private parks, heathland, and mines and quarries in an area that is predominantly rural is very instructive. It is a cardinal error to attempt to show more detail in any map than the scale of the map will carry. The greater detail of the 6-inch surface utilisation maps must be transcribed on the base maps by the process of map-analysis, that is by using separate maps for each category or two or three allied categories. We may thus prepare from the surface utilisation maps an industrial map (factories, etc.), an educational map (schools), a sports map (differentiating public and private sports grounds and kinds of sport, e.g. golf, cricket, football, tennis, race courses, archery, etc.), a map of public open spaces, a footpath map and many others.

A very important branch of survey work is that dealing with agriculture. From the surface utilisation maps it will be possible to construct a map showing the distri-

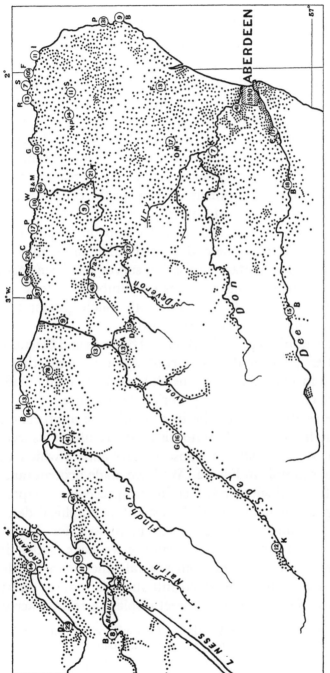

Fig. 20. North-eastern Scotland; distribution of population (1921). Each dot represents 50 people. Towns are shown by circles, the enclosed figures giving their population *in hundreds*.

bution of arable and grassland, rough pasture, market gardens, etc. If the survey area is a small one a crop survey may be made annually, or at least in a given year, and the distribution of crops shown on a base map. This will scarcely be possible for an area containing several parishes unless the co-operation of village schoolmasters is secured. The Agricultural Statistics issued annually by the Ministry of Agriculture give the acreages of the various crops and numbers of stock for the counties only. The figures for separate parishes may be obtained from the Ministry on payment of a fee and on certain conditions, and from these dot-distribution maps may be prepared. If a surface utilisation survey has been made the dots representing acreages or numbers of stock can be placed within the parishes with much greater precision than will otherwise be possible.

In an address entitled "Regional Surveys and Scientific Societies", delivered at the British Association meeting at Oxford in 1926, Sir John Russell has offered many useful suggestions for local survey work in agriculture.[1]

The reader will now be in a position to appreciate a very useful method of summarising the results of a survey which we may call the transect chart and of which we give an example in Fig. 22. We have made acquaintance with the geological transect in the preceding chapter. In Fig. 21 we have added the usual topographical detail to the contour and geological map used for Fig. 12 (p. 81) and to the same transect we have added in Fig. 22, data collected from several other maps in the atlas. We might have added still other data to the chart. Whereas the geological transect displays a section across

[1] Report of the British Association for 1926 and reprinted in *The Geographical Teacher*, vol. XIII, Pt 6, Autumn, 1926.

the geological and contour maps alone, the transect chart may be made to represent a section across the whole atlas or any selected parts of it that we choose. The example we have given shows the features relative to an actual line of section. When the nature of the region permits it is still more effective to construct a generalised transect chart or charts for the whole region or for parts of it that are geologically homogeneous. A specimen of such a generalised chart may be seen in the authors' contribution to *Great Britain*.[1]

We have already alluded to the uses of the transect chart in educational surveys (see pp. 9 and 86). In the class each pupil may build up a chart in successive lessons. It will add greatly to the children's interest in the work if a copy of the geological transect is made as a wall diagram and the other features are shown by mounting in the appropriate places on the chart picture postcards or other illustrations brought in by the children themselves.

We shall conclude the present chapter by giving, in the order of the conspectus, a list of maps, diagrams, tables, etc. that may be prepared or compiled. The list is intended to be suggestive rather than exhaustive. It might be regarded as an index to the portfolio and file cabinets of a well advanced survey. We would again warn those who contemplate embarking upon a survey against entertaining a feeling of discouragement at the apparently overwhelming magnitude of the task before them and perhaps abandoning it in consequence. We would ask beginners, in looking through the following list, not to despair because of the number of suggestions that seem to be beyond their capacity but rather to take encourage-

[1] *Great Britain, Essays in Regional Geography*, edited by Alan G. Ogilvie. Cambridge University Press, 1928.

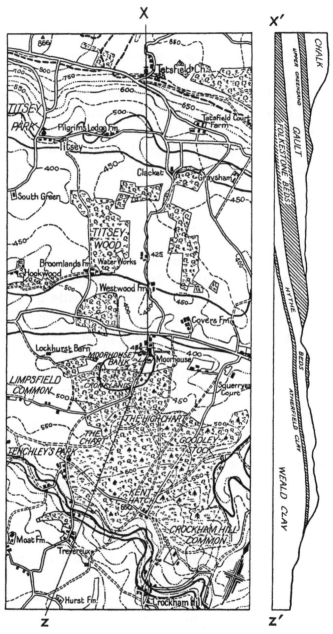

Fig. 21. Geological map and transect, Limpsfield, Surrey. (See Figs. 11 and 12.)

	FARMS				FARMS	PUMPING STATION	FARMS	MOORHOUSE HAMLET AND FARM			ROMAN REMAINS	FARMS
HUMAN SETTLEMENTS												
COMMUNICATIONS	THE OLD ROAD ("PILGRIMS WAY") EAST-WEST			CLACKET LANE				MAIN ROAD EAST-WEST			EAST-WEST ROAD CONNECTING HILL SETTLEMENTS	
MINERAL PRODUCTS				BRICKS AND TILES	SAND					ROAD METAL		
AGRICULTURE	SOME ARABLE	ROUGH GRAZING	ARABLE BELT OATS, RYE ETC. FOR CATTLE		CATTLE AND MILCH COWS	ARABLE BELT	SHEEP MILCH COWS			AGRICULTURAL WASTE		ARABLE BELT HOPS	CATTLE AND MILCH COWS SOME SHEEP
VEGETATION	HAZEL COPPICE WITH OAK STANDARDS	BEECH-YEW WOODLAND CHALK SCRUB & GRASSLAND			DAMP OAKWOODS NEUTRAL GRASSLAND & RUSHES	CONIFERS GRASS-HEATH	AQUATIC VEGETATION		PINES TRUE HEATH	BEECH, CONIFERS, BIRCH			OAK WOODS & NEUTRAL GRASSLAND ALDER WILLOW
NATURAL DRAINAGE	DRY SURFACE CRAY BASIN	SPRINGS	← D A R E N T		OBSEQUENT STREAMS	WELL	SUBSEQUENT STREAM	B A S I N		SECONDARY CONSEQUENT (PARTLY SUBTERRANEAN)	DRY SURFACE → SPRINGS	EDEN (MEDWAY) BASIN OBSEQUENT STREAMS	
RAINFALL	33·0″ — 45·0″ — MAXIMUM ANNUAL — 42·5″	— 25″ MINIMUM ANNUAL — 22·5″	— 32·0″ — 31·0″ — AVERAGE ANNUAL — 30·0″								42·5″ 22·5″	30·0″	
SUBSOIL	FLINTY CLAY	CHALK	SANDY LIMESTONE		C L A Y	SAND & GRAVEL				SANDSTONE & CHERT		C L A Y	
PHYSICAL FEATURES	NORTH DOWNS	CHALK PLATEAU	CHALK ESCARPMENT (BOTLEY HILL)	UPR GREENSAND SECONDARY ESCARPMENT	VALE OF HOLMESDALE		DETACHED SANDY HILLOCKS	MOORHOUSE BANK		DIP SLOPE		GREENSAND RIDGE ESCARPMENT	WEALDEN PLAIN
GEOLOGY	CHALK	CLAY-WITH-FLINTS		UPPER GREENSAND	GAULT		FOLKESTONE BEDS	HYTHE BEDS	FOLKESTONE BEDS OUTLIER	LOWER GREENSAND HYTHE BEDS		ATHERFIELD CLAY	WEALD CLAY

Fig. 22. A specimen transect chart, Limpsfield, Surrey.

ment from the even greater number that are well within their powers. It is true that a few of the suggested maps may need preliminary field work, performed by specialists, spread over a number of years. But these can wait until someone with the ability and inclination comes along to perform the work. It is one of the great advantages of an organised scheme of regional research that whenever a piece of work is accomplished there is a place to fit it in.

Moreover, much that at first sight may seem formidable will become less so as the work proceeds. As the survey workers become accustomed to the use of maps and to finding their way to the sources of information about the region, there will be found to be no limit to the new ideas for making maps and every additional one will be found to throw new light upon many of the others in the collection. If survey organisations associate themselves with the Institute of Sociology, Leplay House, Westminster; the Geographical Association, which has its headquarters at Aberystwyth but has a branch in most large centres; or in Scotland with the Outlook Tower, Edinburgh, they will hear from time to time of opportunities of visiting or contributing to exhibitions of regional survey work, which are always a fruitful source of new ideas and fresh inspiration.

In the list which follows we have given the conspectus numbers using always the cipher for the third figure, the third degree of analysis being left in the hands of local survey groups.

GENERAL RESULTS AND INTERPRETATIONS. ooo

ooo Surface Utilisation Maps (6-inch scale).
ooo Generalised Surface Utilisation Map (scale of base map).
ooo Transect Charts (see Fig. 22, p. 115).
ooo Valley Section Interpretation (see Figs. 24 and 25 (pp. 138, 141).

GEOGRAPHICAL POSITION AND ENVIRONS. 010

010 Quarter-inch map showing the position of the survey area in relation to the surrounding country.
010 Profile of the valley of the river in the basin of which the survey area is situated (see Chapter x).
010 Orientation chart showing directions and distances of principal towns in the country and principal capitals in the world (see Chapter III).

BIBLIOGRAPHY. 020, 030

020 Card indexes of books, maps, papers and other literature relating to the region.
020 Maps showing areas dealt with in guide-books, parish histories, directories, etc.
030 Maps prepared as indexes to the 1-inch, 6-inch and 25-inch Ordnance Survey map issues relating to the region.

GEOLOGY. 100

110 Geological Survey Drift map ($\frac{1}{4}$-inch scale) including and showing position of the survey area.
110 Geological map (Drift) of survey area (1-inch scale or scale of base map).
110 Ditto as uncoloured transparency.
110 Geological map (Solid) of the survey area (1-inch scale or scale of base map).
110 Geological Drift maps (6-inch scale) relating to the survey area.
110 Index map numbering all geological exposures, large or small. Sections no longer visible but of which records are available may be numbered in the same series but in ink of a different colour.
110 Cards with descriptions of geological exposures numbered as in above map (see Fig. 18, p. 102).
110 Index map numbering sites of vertical sections (well-borings, mine shafts, etc.).
110 Sheets of vertical section diagrams numbered as in above map (see Fig. 12, p. 81).
110 Index map showing and numbering lines of geological transects.
110 Sheets of geological transect diagrams numbered as in above map (see Fig. 3 a, p. 41).
120 Map showing distribution of *outcrops* of various types of rocks (e.g. sandstone, clay, limestone, slate, granite) irrespective of stratigraphy.

118 INTRODUCTION TO REGIONAL SURVEYING

120 Map showing outcrops of permeable and impermeable strata.
121/129. Physical and chemical descriptions of rocks.
130 Maps showing distribution of minerals.
130 Descriptive list of minerals.
141/149 Lists of fossils in the various strata, with illustrations.
160 Soil map $\Big\}$ compare with 120.
160 Subsoil map

METEOROLOGY. 200

200 Collection of various weather charts including and showing position of the survey area.
210 Wind-roses and wind-rose maps.
210 Maps showing isobars.
220 Average annual rainfall map (isohyets).
220 Ditto as transparency.
220 Maps showing average monthly rainfall.
220 Isohyets for particular years or months.
220 Storm maps (see Fig. 19, p. 105).
230 Map showing isohels.
230 Map showing aspects (compare with 310).
240 Maps showing isotherms.

RELIEF. 310

310 Contour map with coloured layers (base map scale).
310 Ditto as uncoloured transparency.
310 Maps showing areas above and below significant altitudes.
310 Table of areas at each altitude stage (this may be included as a marginal note on the contour map).
310 Sheets of silhouette transects (see Fig. 16, p. 97).
310 Contour map with numbered index to above.
310 Large-scale (say 3-inch) contour map.
310 Block diagrams (see Fig. 17, p. 99).
310 Relief models (see Chapter VII).

NATURAL DRAINAGE. 320 to 360

320 Small-scale map of river basin including and showing position of survey area.
320 Map showing major and minor catchment areas (watersheds) in the survey area.
320 Map showing rivers, streams, springs, lakes and ponds.
320 Map combining the two preceding.
320 "Regimen" graphs of rivers and streams.

320 Gaugings of flows at different points in rivers and streams at different seasons.
320 Map showing seasonal streams (winterbournes).
330 Drainage map showing also distribution of damp and saturated soils, marshes, etc.
340 Index map numbering sites of wells.
340 Sheets of well sections showing levels of water.
340 Graphs showing fluctuations of water-levels in wells.
340 Map showing underground water contours.
340 Geological transects illustrating underground water phenomena (water-tables, artesian wells, springs).

VEGETATION. 400

400 Map showing areas occupied by natural and semi-natural vegetation.
410/470 Map showing distribution of woodland, heath, grassland, etc. without differentiation of types.
410/470 Vegetation map (advanced ecological study). See for example *The Vegetation of the Peak District*, by C. E. Moss.
410/470 Vegetation transects.
410/430 and 460 Quadrat charts. See *Practical Plant Ecology*, by A. G. Tansley.
480 Maps showing distributions of significant or interesting species.
480 Notes on life histories of species.
490 Floristic lists of selected areas from map 400 above.
490 Flora of the region (marked off in London Catalogue or British Museum Catalogue).

ANIMAL LIFE. 500

510 Maps showing distributions of wild mammals, reptiles, land molluscs, etc.
520 Maps showing distributions of species of birds.
520 Notes on bird migration and territory in bird life.
550 Map showing distribution of stagnant and running water.
550 Pond and river faunas.
580 Notes on life histories of species in relation to environment.

PREHISTORY. 600

610/620 Map showing distribution of plateau and river gravels and alluvial deposits, differentiating by colours deposits of different ages.
610 Map showing sites where alleged eoliths have been found.

620 Map showing sites where Palaeolithic implements and fossil remains have been found, differentiating culture periods.
620 Maps showing hypothetical reconstruction of physical features of region and distribution of population in Palaeolithic times.
620 Notes on the Ice Age in the region.
630 Map showing sites of Neolithic barrows and places where Neolithic implements have been found.
630 Map showing hypothetical reconstruction of physical features (including vegetation) and human settlements in Neolithic times.
640 Map showing places where bronze hoards and Bronze Age objects have been found.
640 Bronze Age barrows and earthworks (including circular churchyards or churchyards that bear evidence of having been originally circular).
640 Hypothetical reconstruction of conditions in the region during the Bronze Age.
650 Hypothetical reconstruction of conditions during the Early Iron Age.
660 Map showing sites of all earthworks, megaliths and roads of pre-Roman date, differentiating periods.
660 Enlarged plans of earthworks, etc.

HISTORICAL SURVEY. 700

700 Map showing sites of Romano-British settlements, Roman villas, etc., places where Roman remains have been found and Roman roads. (See *Victoria County Histories* for symbols.) This map will be strictly a record of known facts.
710 Reconstruction (partly hypothetical) of conditions in the region during the Roman occupation (based upon above map).
710 Enlarged plans of Roman settlements and villas.
720 Map showing Early English settlements.
720 Enlarged plans of village sites.
720/730 Sites of conflicts between Saxons and Normans.
730 Map of region based upon Domesday survey.
730 Extract from Domesday survey relating to the region.
730 Maps showing Norman castles, manors, churches, etc.
730/740 Enlarged plans of mediaeval manors (based upon documentary records, field names, e.g. mill-field, lammas-field, and upon field studies).
730/740 Map showing mediaeval religious life (churches, monasteries, priories, etc.).
720/770 Maps showing existing buildings dating from the various periods.

The above represents only a small selection of the maps that may be prepared to illustrate the history of the region. The same is true for sections 800 and 900 below, each sub-section of which will be treated historically in addition to the above general historical survey. For instance, under 870, a series of maps will show the development of roads, canals, etc. throughout the historic period.

ECONOMIC SURVEY. 800

810 Population (density) map, 1921 census.
810 Ditto 1911 and preceding periods.
810 Population (dot-distribution) map, 1921 census.
810 Ditto as transparency.
810 Ditto for 1911 and preceding periods (with aid of contemporary maps when available).
810 Population increase maps and graphs for intercensal periods.
810 Anthropometric note on inhabitants of the region.
820 Land ownership map.
820 Land tenancy map.
820 Land values map.
830 (Soil maps under 160.)
830 Map showing agricultural utilisation of land in the region (from surface utilisation maps).
830 Map of farms indicating different classes (e.g. dairy, mixed, poultry, etc.).
830 Crop maps (6-inch scale).
830 Maps showing distributions of crops and live stock (dot-distribution).
830 Map showing agricultural markets, storage points, and lines of communication.
830 Map showing areas devoted to timber cultivation, indicating kinds of timber, saw mills and timber yards.
840 Map of mines and quarries indicating produce (coal, iron, building stone, lime, sand, gravel, etc.).
850 Map showing factories, etc. (classified).
850 Map showing factories, etc., using local agricultural products (breweries, flour mills, tanneries, etc.).
850 Map showing factories, etc., using local mineral products (lime kilns, cement works, brick works, potteries, smelting works, etc.).

850 Map showing factories, etc., using raw material, not of local origin (gasworks, chemical works, textile factories, etc.).
860 Map showing engineering works.
860 Map showing contractors yards, timber yards, etc.
870 Small-scale map of region and its environs showing road and railway communications with other towns.
870 Map of region showing railways, roads (classified), and navigable waterways.
870 Silhouette profiles of main roads.
870 Railway accessibility map (see p. 109).
870 Map showing accessibility to other public conveyances.
870 Maps based upon traffic censuses.
870 Communications map (historical).
880 Map showing market places and shopping centres indicating relative importance and areas served by each.
880 Maps showing statutory areas of gas companies, electric supply companies, and water undertakings (with marginal notes as to prices of gas, current, etc.).
880 Postal service map (showing general, branch and sub-post offices, and sorting offices with districts served by each).
880 Map showing telephone exchanges with areas and in rural areas wiring systems.
890 Map showing distribution of various banks, insurance offices, etc.

SOCIAL SURVEY. 900

910 Maps showing the distribution of persons engaged in the various occupations. (Compiled from census and directories.)
920 Large-scale maps showing types of houses in urban areas: residential, hotels, flats, artisans houses, slums and tenements.
920 Historical housing map showing by colours houses built at various periods up to and including post-war. (See also 720/770.)
920 Hospitals, asylums, etc.
920 Various maps showing vital statistics by dot-distribution method.
930 Civil administration (boroughs, urban and rural districts, parishes, poor-law unions, town halls, etc.
930 Parliamentary divisions with supplementary historical and statistical data.
940 Map showing military and naval establishments.
950 Education authorities' areas.
950 Map showing distribution of schools of various grades.
960 Public open spaces and playing fields.
960 Private sports grounds.
960 Golf courses.

960 Indoor clubs and friendly societies, classified.
960 Public-houses, classified.
960 Theatres, music halls, cinemas, etc.
970 Libraries, public and private.
970 Large-scale maps of place-names, field-names, public-house signs, etc.
970 Card index of names with etymological notes.
970 Local customs, folk-lore, etc.
980 Ecclesiastical parishes with churches; diocese, rural deaneries.
980 Sectarian religious houses.

CHAPTER IX

PICTORIAL ILLUSTRATIONS

THERE is in a regional survey great scope for the photographer, for there is scarcely a map that can be made or a subject treated which does not lend itself to pictorial illustration. In many cases, as for instance geology, vegetation, or architecture, the scope for photography is almost unlimited. It will sometimes be useful to mount one or two small prints on unused portions or on the margins of the maps themselves, but the general survey collection of illustrations are, in our opinion, best mounted upon cards of a dark neutral tint, with a neat, preferably printed, label at the top left-hand corner. The practice of inserting photographs in albums is not recommended. The card mount system is much more convenient in every way and it enables us to insert new illustrations in the cabinet, as they come along, in their proper places according to the order of the conspectus.

A very suitable size of mount is that adopted by the County Photographic Survey and Record Societies, namely $13'' \times 10\frac{1}{2}''$. This is large enough to take whole plate prints but as a general rule half-plate is the size that should be aimed at, if necessary by enlargement from smaller negatives. The co-operation of a local photographic society will be very useful, both for supplying prints and making enlargements. Prints of a size smaller than quarter-plate are generally unsatisfactory for this method of mounting. If, however, wholesale enlargement is not practicable they may find a place in a survey as mentioned below.

The frontispiece is a reduced representation of one of the photographic illustrations of the Croydon Regional Survey. This particular print has been placed under the heading of relief (310), but it might equally well have been used to illustrate geology (110), vegetation (430), agriculture (830), history (740), or communications (870) for the path in the foreground is a portion of the Pilgrim's Way. Many of the illustrations will have a manifold interest such as this. In some instances it is very desirable to insert duplicate prints in the collection under different subjects, but in any case a system of cross references is useful.

A brief description, stating the points of interest it is intended to illustrate, should be added to each photograph. If the space on the label is insufficient for this purpose a neat strip of paper may be fixed to the bottom of the mount and the description written on it and, if necessary, continued on the back of the mount.

In some surveys the practice has been adopted of mounting small photographs, say $3\frac{1}{4}'' \times 2\frac{1}{4}''$, postcards, sketches, etc., on white foolscap sheets and writing on the same sheets a full description of the picture. This affords an easier method of dealing with small photographs than that described above and may be followed when enlargement from the small negatives presents difficulties. It is desirable, however, in the authors' opinion, to add aesthetic quality to the survey records when this can be achieved without undue absorption of time and resources and without detriment to their primary purpose as records.

The whole subject of photography for the purpose of survey records—apparatus, subjects, processes, mounting, classification, labels, and filing—has been fully discussed in a book published in 1916 entitled *The Camera as Historian*, by Messrs Gower, Jast and Topley, three pioneer members of the Photographic Survey and Record of Surrey. This work is well worthy of study by all regional survey photographers. The system of labelling there described, however, is more elaborate than is needed for purposes of regional survey. Nor does the classification of subjects adopted, based upon the Dewey system, suit our purpose so well as that set out in the conspectus in this volume.

It is of the greatest importance that prints should be made by a process that gives permanent results. The authors of the above book advocate the platinotype and carbon processes which certainly give excellent results both as to effect and permanence. The simpler and less expensive process of sulphur toned gaslight or bromide prints is more within the scope of the average photographer and gives results that are equally permanent and very pleasing. For the benefit of photographers who have not given consideration to this aspect of the subject we have appended to this chapter a short paper by Messrs E. A. Robins and J. H. Pledge in which are given instructions and formulae for making sulphur toned prints.

The pictorial illustrations of a survey need not be confined to photographs. Engravings from old books depicting scenes or objects relating to the survey area should as far as possible be collected and mounted on cards. These can often be picked up cheaply on book stalls and it is the experience of teachers that school children will produce them abundantly to illustrate a school survey. Old engravings often depict scenes that are no longer available for the photographer. When such, owing to their scarcity, cannot be acquired for the survey collection, access can usually be had to them for copying by photography or otherwise. The same is true of old and scarce maps.

Picture postcards are not to be despised. They are particularly useful in school surveys to which they may be contributed by the children. They often furnish a survey with admirable illustrations and sometimes such as cannot be otherwise obtained. Aerial photographs of parts of the region are of special interest but aerial photography is not yet within the reach of the majority of regional surveyors. Whenever postcards or larger prints of aerial photographs are obtainable they should be added to the survey collection.

A valuable adjunct to a regional survey is a collection of lantern slides. These are very easily made by contact with small negatives or by reduction from larger ones. Slides of the maps and diagrams should be included in the collection. For this purpose it is necessary to copy them, using process plates, before they have been coloured and when necessary to colour the slides afterwards by hand. There are various colours on the market that may be used but the most brilliant results on the screen will be obtained by the use of oil colours. Of these crimson lake, Italian pink, Prussian blue, verdigris, and alizarin yellow are absolutely transparent and burnt sienna and burnt umber nearly so. A wide range of colours may be obtained by mixing these. The colours are best mixed with copal varnish and camel hair brushes are quite satisfactory for their application to the slides.

Diagram slides may be made by hand with drawing ink on what are known as notice plates. These can be purchased cheaply and they may be coloured in the same way as ordinary lantern slides.

It is very convenient to use white binding for one edge of the finished slide, the edge above the spots. The short title of the slide and its conspectus index number can be written on this edge and the slides kept in order in the cabinet with their titles uppermost. A card index to the slides giving the facts and points of interest about the slides is also very useful.

The acquisition of a lantern slide collection, while not essential to a survey, makes possible its demonstration in a popular way to large audiences, thereby adding to the public interest in the work.

We may here appropriately give some consideration to the custody and exhibition of survey material. The maps and other material forming the regional survey atlas may be kept in a portfolio or still better in a loose-leaf cover. The latter will probably need to be specially made for the purpose if the sheets are larger than quarto size. In the case of a school survey the accumulating material is usually kept at the school but it is desirable from every point of view that the results of a survey society's work should be made available for public reference. The most appropriate place for this will usually be the public library or museum, or in rural centres the village institute. At the library the survey material will take its place as a valuable addition to the local collection which is the pride of most librarians. A selection of the maps may with advantage be exhibited from time to time or duplicates of some of them framed and hung permanently upon the walls.

In some centres there already exist museums devoted to the natural and human history of the town and its environs. Such institutions are the ideal headquarters of a regional survey society and when they are not in being their foundation may well be one of the society's aims. In such a home there is unlimited scope for a survey organisation to make its activities felt as an educational influence of the highest value.

MAKING PHOTOGRAPHIC PRINTS FOR REGIONAL SURVEY[1]

By E. A. ROBINS

AND

J. H. PLEDGE

The object of the present paper is to lay down a few simple rules for the production of photographic prints of a high degree of permanence. Judging from the photographs we have seen at recent exhibitions of Regional Survey work it appears to us to be very necessary to direct the attention of regional survey photographers to this matter, for few of the prints examined will be available for future reference owing to the fugitive character of the processes used or want of sufficient care in carrying it through.

We need not here discuss the making of negatives as the final print is the subject of our present consideration. The permanence or otherwise of a photograph depends upon a number of factors. For example, the process selected for the production of the prints, the method of mounting, the susceptibility of the print to the dampness of the atmosphere and so on.

THE PROCESS

There are many photographic processes for producing prints, for example, P.O.P., collodio-chloride, gaslight, and bromide, among the processes in which silver is the image producing substance, platinotype, in which it is metallic platinum, and the carbon process in which an insert pigment is used.

Of these, the P.O.P. and collodio process can be ruled out as they give decidedly fugitive images, even when produced under the best conditions of toning and fixing. We fear that many of the above-mentioned prints have been made by printing on P.O.P. and toning with a combined toning and fixing bath—all such prints are totally unsuitable for preservation. Gaslight

[1] Reprinted from *South-Eastern Naturalist* for 1925, by the courtesy of the authors and of the Council of the South-Eastern Union of Scientific Societies.

and bromide prints are much more permanent, and although slightly more difficult to handle, it seems to us that it is worth taking extra trouble in order to make records that will be permanent.

THE PRODUCTION OF THE PRINT

Either gaslight or bromide prints can be made absolutely permanent so far as fading is concerned, by toning in a sulphide toning bath, and thus converting the silver image into one of silver sulphide of a pleasing brown colour. There are two methods of doing this. The hypo alum toning bath is preferable, but slightly the more difficult owing to the toning bath having to be kept at a temperature of about 130° F. during the process. This gives results which may be considered as absolutely permanent, at least so far as the image is concerned, and can be applied to prints produced on both gaslight and bromide papers. The other is a cold sulphide toning process in which the print is bleached in a special bath, and after washing, is toned brown in another solution, producing a silver sulphide image. This process is easily accomplished and gives excellent results, which, so far as we know, are quite permanent, not being changed by time, light or moisture, and we strongly recommend that all prints be produced by this method.

In order that the best results may be obtained, the photographic paper makers' instructions should be followed closely, and especial care be taken that the print is not over or under exposed, the exposure should be such that after about $1-1\frac{1}{4}$ min. development, the image has appeared to its full depth, and no more would be produced by longer development. Great care must be taken in fixing, and the use of successive baths is recommended to make quite sure that all the soluble silver is removed. From the point of view of permanency perfect fixation is far more important than prolonged washing, although, if toning is the ultimate object, the washing must be very thorough, say at least 20 min. in running water. A formula for a developer is given which is generally applicable to all papers, but we strongly advise the amateur to keep to one kind of paper, and use the developer recommended by the maker.

PICTORIAL ILLUSTRATIONS

Glossy bromide, or gaslight should always be used for record prints, as fine detail is the main object to be aimed at, rather than artistic effect. The toning is easily carried out by bleaching in a mixture of pot. bromide and pot. ferricyanide, until practically all traces of the image have disappeared, then washing in water until the yellow colour is washed out, and finally, placing in a bath of sodium sulphide until the image is toned. A short wash completes the operation. In order to obtain the best results the glossy prints should be dried on ferrotype plates: this is easily accomplished by squeegeeing them on to ferrotype plates that have been rubbed over with common benzole. The prints dry rapidly and come off with a good gloss.

Among other photographic processes suitable for record prints must be mentioned platinotype, which gives very beautiful permanent results, but which, unfortunately, is not very often used on account of its high price. Only matt surfaces are possible by this process. There is also the carbon process in its original form, or in the modified form known as carbro, placed on the market by the Autotype Co. Either of these entails rather more detail work, but gives permanent results that are very suitable for preservation.

Attention should be paid to the mounting of the prints, and where possible, dry mounting tissue should be used in place of flour paste or any similar method of mounting. The latter are liable to set up fermentations, and produce acid results, with a consequent destructive effect on the prints. Dry mounting is carried out by the use of a thin paper impregnated with shellac, which is placed between the mount and the print, and the two caused to adhere by pressing with a hot iron. The process is easy, and quite insulates the print from the mount.

FORMULAE USED FOR BROMIDE AND GASLIGHT PAPERS

All prints developed in:

Metol	1 gm.	30 grains
Hydroquinone ...	4 gm.	120 grains
Sod. sulphite cryst. ...	40 gm.	3 oz.
Sod. carbonate cryst.	40 gm.	3 oz.
Pot. bromide 10 % sol.	1·5 c.c.	80 drops
Water to	1000 c.c.	80 oz.

and fixed in:

Hypo	...	...	200 gm.	1 lb.
Water	...	...	1000 gm.	80 oz.

to which was added:

Sod. sulphite cryst.	...		12·5 gm.	1 oz.
Acetic acid (glacial)	...		9·5 c.c.	$\frac{3}{4}$ oz.
Alum	...	...	12·5 gm.	1 oz.
Water to	...	...	125 c.c.	10 oz.

After fixing and thorough washing the prints are placed in the bleaching bath until all the black image has disappeared, they are then washed until all the yellow stain has gone and placed in the toning bath until the image has turned a sepia colour. A final good washing completes the process.

FOR SULPHIDE TONES

Bleach in:

Pot. ferricyanide	...		25 gm.	$\frac{1}{2}$ oz.
Pot. bromide	...	...	25 gm.	$\frac{1}{2}$ oz.
Water to	...	...	500 c.c.	10 oz.

Tone in:

Sod. sulphide	...	...	5 gm.	44 grains
Water to	...	...	500 c.c.	10 oz.

HYPO ALUM TONING BATH

Hypo	...	...	1 lb.
Water	...	...	80 oz.

Add:

Alum	...	...	$2\frac{1}{2}$ oz.

Boil and cool to 140° F.

Prints are washed and dried after fixation, and then placed in the preceding bath, heated to 130–140° F. until the desired tone is attained. Further washing and drying completes the process.

This bath can be used repeatedly and improves with age. An enamelled iron pie dish is suitable for containing it while heating and in use.

CHAPTER X

INTERPRETATIONS AND APPLICATIONS

AT the close of a work on the technique and methods of regional surveying we may appropriately ask what are its wider significance and utility. In its broader aspects we may regard the regional survey movement as an attempt to establish a firm observational foundation upon which to build a science of Sociology. Our individual efforts in observing human communities in their environmental settings and in recording our observations are no less directed towards this end because we may not be fully conscious of our aims. Out of the growing body of material resulting from regional researches we may legitimately expect that new interpretations of social phenomena will arise and react beneficially upon educational and social policy. We should not, however, strain after interpretation. If we patiently carry on our field observation and produce our maps and other records or expressions of our work, the relationships of cause and effect between the varied phenomena of the region will gradually unfold themselves to our vision and something of the meaning of the regional drama will dawn upon us. Whatever our individual interests may be we shall find they have contacts with every other aspect of the region in the manner broadly indicated in the conspectus diagram.

In order to help us in our understanding of the region, how it has come to be what it is, how it is changing before our eyes and what it may become, we shall read, concurrently with our field studies, all that has been written about it or as much as time and opportunity will permit. Guide books, local histories, papers buried in the publications of learned societies, all are grist to the mill of the regional surveyor. Above all we shall not neglect good works of fiction that have their settings in or about our region: Hardy (Wessex), Kipling and Sheila Kaye-Smith (Sussex), Eden Phillpotts (Devon), Marriott (Cornwall), Mrs Gaskell (Lancashire, Cheshire), the Brontë Sisters (Yorkshire), Dickens (Kent and many other places); the list might be indefinitely extended. Such authors have a keen regional sense and a sympathetic insight that is not given to many. We recall, for instance, an opening passage in *Great Expectations*—"Ours was the marsh country down by the river". How pregnant are these few simple words for all that follows in the story. Already we have a vision of the place and the folk about which we are to read.

Our field studies will soon give us a new interest, too, in more general treatises on the sciences and the humanities. When we have seen the different rocks exposed in quarries and railway cuttings we shall wish to know something of what geologists have to tell us about them. When we have thought of the village community that was the beginning of our own parish or town in far-off times, and have attempted with the aid of field names and such other data as we can collect to reconstruct our own mediaeval manor, we shall turn with zest to the works of Maitland and Rowntree, Gomme and Harold Peake, and not without some power of estimating the value of their various interpretations. Again, when we have noted the different types of men and women collected together in our town we shall wish to know something of anthropology, or when we have learnt to observe survivals of the past in the buildings and institutions that surround us we shall find new interest in history and architecture.

The field of Sociology, in spite of voluminous literature, is still very largely an open one. As an observational science, Sociology can scarcely yet be said to exist. It can only be placed upon a sound footing beside the other field sciences as a result of continued regional research. The Newton or Darwin of Sociology has still to make his appearance, but it is safe to predict that he will emerge, if at all, from the ranks of regional survey workers, or at least he will build upon the results of their labours. His precursors, however, may already be discerned and of these the foremost place must be given to Professor Patrick Geddes, who has been the main inspiration of the modern regional survey movement. Not only has he given direction and imparted enthusiasm to two generations of workers in diverse fields but he has himself led the way in the sociological interpretation of regional phenomena.[1]

Professor Geddes in turn has taken much from the French sociologists and particularly from Frederick Le Play (1806–1882), but he has greatly developed their work and has added much of his own. Le Play corrected

[1] There is unfortunately no satisfactory exposition of Geddes' work in regional interpretation in a convenient form. This defect we hope to remove by a later volume. Even a bibliography of his scattered papers and those of his colleague, Mr Victor Branford, in the *Sociological Review* and other journals, in the form of reports on town and regional planning, in encyclopaedia articles and in books, would be too long for insertion here. Pending such a publication, readers may write for guidance to the secretary of Leplay House, 65 Belgrave Road, London, S.W. 1, or to the Outlook Tower, Castle Hill, Edinburgh. Happily the inaccessibility of literature need trouble no serious student to-day. An application through a local librarian to the Central Library for Students for the loan of any book rarely meets with disappointment.

the utilitarian economists of his day by pointing out the simple truth that mines produce not merely coal and ores but, even more significantly for civilisation, a human type, namely miners; pastures not merely sheep, but shepherds also and so on for all the human occupations. Each type has its characteristic outlook upon life and has made its specific contributions to culture and the social milieu. Members of the school of "La Science Sociale" (founded by Demolins and De Tourville, following Le Play) worked out in a magnificent series of

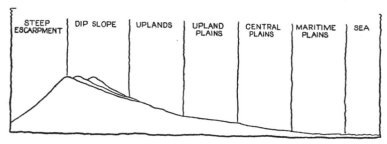

Fig. 23. The zones of the valley section.

monographic studies in different parts of Europe the interactions of family and place.

Professor Geddes, from a less purely economic standpoint, has somewhat modified Le Play's classification of rustic types and has given us a synthetic view of them all placed side by side in their positions on the "valley section". This is the most helpful key that has yet been fashioned for the elucidation of human geography. We cannot here give an adequate account of the "valley section" and all that Professor Geddes has built upon it. We must content ourselves with a brief exposition of the elementary idea. The diagram (Fig. 23) is a generalised representation of the most typical form of land mass to be found in all parts of the world. It might represent the profile of England and Wales from the mountains in the west to the lowland plains in the south-east—the combined valleys of the upper Severn and the Thames which as geologists tell us were formerly parts of one and the same extensive drainage basin. Still more clearly might it be the profile of one of the Scottish river valleys or those of Scandinavia from the Norwegian mountains to the Baltic seaboard. Again it might be taken as representing the section from the Alps to the Baltic, the Alps to the Mediterranean or the Andes to the Atlantic. In short it is the generalised section of any well-developed river valley.

We see that it is capable of division into seven or eight zones or regions. Beginning on the left of the figure we have first the steep and rocky escarpment. Next are the highlands of the gentler dip slope followed by uplands and again by upland plains, often deeply dissected. Then the broader central plains and beyond them the fertile alluvial plains of the maritime belt and finally the seacoast and the sea.

Broadly speaking each of the zones is the habitat of one of the eight rustic types of humanity distinguished by Geddes. As cultural types these have been created by their several environments with differing natural resources and the occupations proper to them. On the other hand man spends activity upon his environment, exploits or develops it to satisfy his needs and so far dominates it in turn.

The distribution of the rustic types over the valley section is indicated in Fig. 24. On the escarpment characteristically is the miner with his pick; on the higher mountain slopes the woodman with his axe and at the

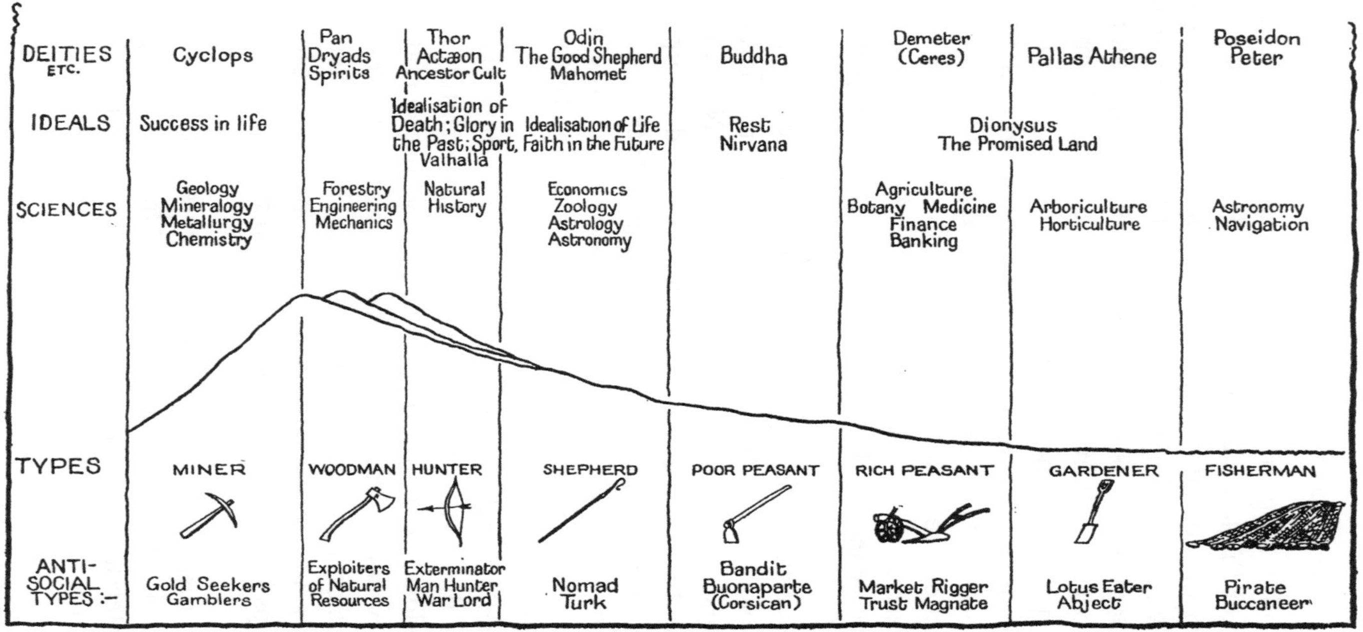

DEITIES ETC.	Cyclops	Pan Dryads Spirits	Thor Actæon Ancestor Cult	Odin The Good Shepherd Mahomet	Buddha	Demeter (Ceres)	Dionysus	Pallas Athene	Poseidon Peter
IDEALS	Success in life		Idealisation of Death; Glory in the Past; Sport, Valhalla	Idealisation of Life Faith in the Future	Rest Nirvana		The Promised Land		
SCIENCES	Geology Mineralogy Metallurgy Chemistry	Forestry Engineering Mechanics	Natural History	Economics Zoology Astrology Astronomy		Agriculture Botany Medicine Finance Banking		Arboriculture Horticulture	Astronomy Navigation
TYPES	MINER	WOODMAN	HUNTER	SHEPHERD	POOR PEASANT	RICH PEASANT		GARDENER	FISHERMAN
ANTI-SOCIAL TYPES :-	Gold Seekers Gamblers	Exploiters of Natural Resources	Exterminator Man Hunter War Lord	Nomad Turk	Bandit Buonaparte (Corsican)	Market Rigger Trust Magnate		Lotus Eater Abject	Pirate Buccaneer

Fig. 24. The valley section with rustic types.

forest edge the hunter with bow and arrow. The uplands provide the natural home of the shepherd with pastoral crook and the upland plains that of the poor peasant or crofter with mattock. The central plains furnish the broad arable tracts ploughed by the rich peasant while the fertile maritime plains are the home *par excellence* of the gardener with his spade. Finally the sea-coast and the sea are the home and workplace of the fisherman plying his net.

The natural environments and consequent occupations of each of these primitive rustic types are reflected in the cultures which they have severally developed. Each has its characteristic sciences, arts and crafts, aims, ideals and even religions, and each has contributed its own particular anti-social representatives to the social order—gold seekers, exterminators and man-hunters, nomads, bandits, financiers, abjects and pirates follow in succession as we pass from left to right. The conductor of an orchestra beats time to instruments that have had their origin in all zones of the valley section just as the mayor keeps harmony or tries so to do between the discordant political representatives who compose the corporations in our towns. The towns have sprung up on the rivers for interchange of products just at those points on the section where two zones meet. Such towns will be found to be dominated by urbanised representatives of the two adjacent rustic types. On the maritime plains at the river mouths the seaports have grown up in which we may see congregated members of every type from every regional zone.

This in barest outline is the view of the regional drama which has been elaborated. It gives us a naturalistic outlook upon which to base our first attempts at regional interpretation. It is for us to try it out in our several regions. We may begin by asking ourselves at what point in the valley section our parish or larger area is situated and then, with the knowledge we have gained of its physical features and natural resources, to see how these have influenced the occupations and cultural development of its folk. If we are dealing with a town or city we shall enquire into the ways in which it reflects its immediate rustic environment on either side and what are its contacts with more remote zones. We shall learn how it is that the town has arisen just at that place and how to detect the rustic types which, in their urban disguises, make up its population and impart to it those characteristics which distinguish it from other towns. It is, of course, only in wide continental tracts that we may observe whole peoples exhibiting the pure culture of one or other of the rustic types—mining, hunting, pastoral and so forth. But by studying these in literature we shall gain much that will illuminate the smaller units as they exist in a minutely varied and highly developed country like our own, where fragments of each zone are scattered over tracts that primarily belong to one. For instance a chalk quarry or clay pit occurring on the maritime plain is to be regarded as the displacement of a little piece of the escarpment, an outlier as we should say in geology. As civilisation advances and man gains increasing powers for the exploitation of natural resources this obscuration of the primitive zones by displacements from others becomes more and more complete, as witness the recent migration of Welsh miners to East Kent, from one end of the valley section to the other.

And yet how well the valley section may still serve our purpose even in a small tributary valley and in close proximity to the metropolis has been illustrated by the Croydon survey. Fig. 25 is a generalised profile of the valley of the Wandle from the neighbourhood of Reigate and Merstham at the foot of the chalk escarpment to Mitcham and Croydon on the "maritime" plain. Considered in relation to the Severn-Thames valley section the whole of the Croydon survey area falls within the central plain and maritime plain zones, and yet within these zones the small valley section of the Wandle basin presents an almost complete epitome of all the types and in their correct sequence on the section. On the escarp-

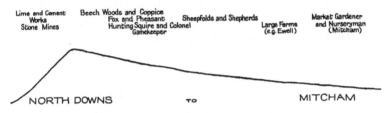

Fig. 25. The valley section, Wandle Basin.

ment at Merstham and Reigate are the extensive mines for firestone and hearthstone and, as also at many similar situations in neighbouring valley sections (Mole, Darent, Medway), great chalk quarries with their lime-kilns and small groups of quarrymen's cottages. Behind the escarpment on the clay-capped chalk plateau woodlands abound often in the form of coppice—with oak standards now chiefly used as game preserves for the sporting squire-colonel, a modern representative of the ancient hunter, though his game-keeper is truer to the primitive occupational type. A little lower on the dip slope sheep-folds are characteristic though less so than in corresponding zones of adjacent valley sections. In the next lower zone where the chalk has a good covering of soil were large arable farms which, in this district, have latterly given way to the advance of the suburban builder. Finally on the alluvial levels around Mitcham the market gardener and nurseryman still flourish, though not to their former extent when Mitcham was renowned for its herb gardens and Mitcham lavender-water was distilled from plants cultivated for the purpose within the parish. In this particular valley section a sea of population takes the place of the maritime zone, but the corresponding section for the Medway or the Stour in Kent would end in the orthodox way with communities of fisherfolk.

We have already indicated in these pages some of the practical applications of regional survey and we have seen that these relate to the problems of the educator on the one hand and to those of the town planner on the other. We may here recall what was said of town planning in the opening chapter, namely that it is to be understood as "the direction of all development so that all land may be put to the use for which it is best fitted, the health, wealth and happiness of the community being paramount considerations".

It is not within the scope of this Introduction to discuss at length the technical aspects of the applications of regional survey to education or town and regional planning. These must be reserved for future volumes. In regard to education, however, we may point out that regional survey is not to be regarded as a new subject clamouring for a place in an already overloaded curriculum. It is rather an attitude of mind which opens up a new method of approach to almost all the subjects that are taught in our schools. It is in fact a systematic application of the golden rule in education by which the pupil is led by gradual steps from the known to the unknown, from the familiar phenomena of his immediate environment to their analogues in the wider realm of general knowledge. We may add that the now numerous experiments that have been made in the application of regional survey to both primary and secondary education have fully demonstrated the value of the method.

We are not among those enthusiasts who would claim for regional survey that it is a panacea for all the social evils. On the contrary we regard the regional survey movement as a symptom rather than a cause of the general striving after social betterment that characterises our times. At its best it is the scientific and technical expression of that striving. We may instance the rise of the medical officer of health, who is pre-eminently a regional surveyor in his most important field, and whose ideal is the healthiest possible community.

Regional survey is thus seen to be a method of organised research into the nature of human communities and human environments and the functional relationships between them. A penetrating historian will perhaps discover that regional survey had been called into existence by the needs of our modern situation. When we reflect upon the triumphs that have already attended the application of the results of scientific research in the realms of physics, chemistry and biology we may reasonably anticipate the emergence of a better social order when the results of scientific research in regional sociology can be applied to the arts of citizenship and administration.

INDEX

Administrative areas, 22, 26, 48
Administrative map, 48, 109
Aerial photographs, 127
Agricultural statistics, 26, 112
Agriculture, 20, 101, 110, 112, 121, 125
  on transect chart, 115
Agriculture, Ministry of, 112
Air Ministry, 104
Alluvial deposits, 119
Altimeter, 89
Animal ecology, 19, 106
Animal life, 10, 14, 17, 19, 106, 119
Anthropology, 10, 15, 134
Anthropometry, 20, 121
Aquatic vegetation, 19
Archaeological literature, 29
Archaeology, 10, 15, 29, 107
Architecture, 107, 124, 135
Area, choice of, 21
Area of survey region, 26
Artesian wells, 119
Aspect (climatic), 118
Atherfield Clay, 80, 81, 114, 115
Atlas of the region, 93–123
Atmospheric phenomena, 18, 103

Bacon, Roger, 8
Base maps, specification, 34–36
Bathymetry, 19
Bibliography, 117
Birds, 106, 119
Block diagrams, 87, 93, 98, 99, 118
Bogs, 19
Books, 51, 134
  on maps, 50
Botany, 19, 119
Branford, Victor, 135
Brighton, 31
Bristol, 32
British Museum Catalogue of Plants, 119
Bronze Age, 19, 120

Card index, 71, 117, 123, 128
Catalogues of Ordnance Survey maps, 50
Census, 26, 108, 109, 121, 122
Central Library for Students, 135
Chalk, 70, 72, 74, 80–82, 99, 114, 115, 141
Chalk escarpment, 99, 115, 141
Chapman, E. H., 103
Civic centre of a region, 25, 26
Civic surveys, 76, 122
Civil administration, 122
Civil parish as unit of area, 29, 34, 42
Clark, J. E., 104
Clay-with-flints, 72, 74, 81
Climate, 10–13, 30, 103, 104, 118
Climatic factors and vegetation, 10, 13, 14
Coal, 121, 136
Coast erosion, 12
Colours, for lantern slides, 127
  for maps, 55, 96, 100
Communications, 20, 97, 121, 122, 125
  on transect chart, 115
Consequent streams, 99, 115
Conspectus diagram, 10
Conspectus of regional survey, 8–20, 22, 93, 106, 116, 126
Copyright of Ordnance Survey maps, 36, 43, 47
Counties as regional survey areas, 29
County Administrative map, 48, 109
County Directory, 109
County Histories, 29
Cox, Rev. J. Charles, 108
Cretaceous System, 82
Crop surveys, 53, 95, 112, 121
Croydon, 31, 75, 141
Croydon Regional Survey, 101, 102, 125, 141

146 INDEX

Damp soils, 119
Darent River, 141
Denudation, 10–12
Dewey system, 15, 16
Dip of strata, 84, 85
Directories, 29, 109, 117, 122
Dissection of maps, 49, 58
Distributive industries, 20
Domesday survey, 120
Domestic architecture, 107
Dot-distribution maps, 109, 110, 121, 122
Dover, 31
Downe, Kent, 67–69, 72, 73–76
Drainage map, 119
"Drift" maps, 100, 117
Dry valleys, 99
Dundee, 32
Dunes, 19

Early Iron Age, 19, 120
Earthworks, prehistoric, 19, 64, 93, 120
East Kent, 140
Ecclesiastical administrative areas, 29, 123
Ecclesiastical organisation, 20
Ecclesiastical parishes, 123
Ecology, 19, 104, 106, 119
Economic geology, 18
Economic survey, 20, 121
Economics, 15, 20, 108, 121
Edaphic factors and vegetation, 10, 13, 14
Edinburgh, x, 116, 135
Education, 3, 20, 122
Electricity supply, 122
Engineering and building, 20
Engravings, 126
Environs of the region, 18, 30–33, 117
Eoliths, 119
Erosion, 12, 19
Escarpment, 115, 140, 141
Exhibitions, 116, 128

Factories, 121, 122
Farms, 121

Farquharson, Alexander, xi
Faulting of strata, 86
Fauna, 19, 119
Fawcett, C. B., 29
Fen, 19
Fiction in regional literature, 134
Field names, 48, 70, 120, 123, 134
Finance, 20
Fishery, 20
Fleure, H. J.
Flora of region, 19, 29, 104, 119
Flow of rivers, 119
Folkestone Beds, 80–82, 85, 114, 115
Folk-lore, 123
Fordham, Sir H. G., 34, 50
Forestry, 20
Form lines, 89
Fossils, 11, 18, 118, 120
Freshwater marsh, 19

Games, 20
Gas companies' areas, 122
Gault, 80–82, 85, 99, 114, 115
Geddes, Prof. P., 135–137
Geographical Association, x, 37, 116
Geographical Journal, 24
Geographical position and environs of region, 18, 30–33, 117
Geographical Teacher, 112
Geological exposures, 56, 60, 64, 70, 72, 84, 101–103, 117
Geological maps, 18, 39–44, 50, 66, 77, 80–83, 100, 105, 117
Geological sections, 41, 82, 117
Geological Society, Journal, 86
Geological Survey, H.M., 39, 66
  colour convention for maps, 100
  maps, 86, 117
  Memoirs, 50, 73, 82, 86, 100, 103, 104
Geological transects, 77–87, 112, 117, 119
Geologists' Association, Proceedings, 86
Geology, 9–12, 18, 80–87, 100, 117, 124, 125, 140
Geometrical outlines of survey areas, 22, 23

INDEX

Government and administration, 20
Graphs, population, 121
Great Malvern, 32
Greensand escarpment, 115
Guide books, 117, 134

Hailstorm map, 104, 105
Hastings Beds, 99
Hawkins, H. L., 88
Heaths, 19, 57, 104, 119
Herbertson, Prof., 21
Hindhead, 32
Hinks, A. R., 50
Historical survey, 19, 120
History, 10, 15, 17, 19, 107, 120, 125, 135
Holmesdale, 99, 115
Horizontal sections, 41, 82
Housing survey, 20, 108, 122
Human region, 23
Humidity, 18
Hydrographic maps, 31, 118, 119
Hydrography, 19, 98
Hythe Beds, 80–82, 85, 114, 115

Ice Age, 120
Impermeable strata, 118
Index maps, 117
Institute of Sociology, x, 116
Interpretations of surveys, 116, 133
Iron Age, 120
Isle of Wight, 83
Isobars, 118
Isohels, 118
Isohyets, 104, 118
Isotherms, 118

Kelly's *Directory*, 109

Laborde, E. D., 50
Lakes, 19, 118
Land ownership, 20, 121
Land tenure, 20, 121
Land utilisation, 20
Land value, 20, 121
Language, 20
Lantern slides, 16, 127, 128
La Science Sociale, 136

Layered map, 118
Layered models, 88
Le Play, Frederick, 135, 136
Leplay House, x, 116, 135
Limpsfield, Surrey—map, 114
Literary associations of region, 134
Live stock maps, 121
Lobeck, A. K., 98
Local administrative areas, 26, 29, 48, 109
London Catalogue of Plants, 119
London School of Economics, 25
Lower Greensand, 82, 83, 99, 115

Manufacturing industries, 20
Map analysis, 95, 108, 110
Marine vegetation, 19
Maritime areas, 18
Markets, 20, 121
Marshes, 19, 57, 119
Mediaeval manors, 120, 134
Mediaeval survivals, 107
Medway valley, 97, 141, 142
Megaliths, 19, 120
Merstham, Surrey, 141
Meteorological Office, 104
Meteorology, 10, 11, 18, 103, 118
Military and naval institutions, 56, 122
Military and naval organisation, 20
Mill, H. R., 24, 25, 30, 93
Mineral industries, 101
Mineralogy, 18
Minerals, 118
Miners, 136, 137, 140
Mining and quarrying, 20
Ministry of Agriculture, 112
Mitcham, Surrey, 141, 142
Mole chalk gap, 90, 91
Mole River, Surrey, 141
Moorland and heath, 19
Moss, C. E., 119
Mounts for photographs, 124
Museums, 128

Natural drainage, 10, 12, 13, 19, 118
 on transect chart, 115

Natural regions, 22
Natural vegetation, 105, 119
Naval and military institutions, 56, 122
Naval and military organisation, 20
Neolithic antiquities, 107, 120
Norman antiquities, 120
North Downs, 115

Obsequent streams, 99, 115
Occupations, 20, 122
Ogilvie, A. G., 99, 113
Omnibus routes, 109
Ordnance Survey maps, 17, 18, 25, 34–50
  early editions, 39, 47
  index sheets, 48
  reproduction of, 36, 43
  special rates of purchase, 36, 39, 49
Orientation chart, 31–33, 117
Orography, 19
Outcrop, 80
Outlier, 80, 85, 140
Outlook Tower, Edinburgh, x, 116, 135
Oxford School of Geography, 25

Palaeolithic implements, 120
Palaeolithic sites, 120
Palaeontology, 18
Parish as a regional survey area, 26–29
Parish history, 108, 117
Parish index map, 108
Parish statistics, 61–63, 75, 76
Parish surveys, 63, 65–76
Parliamentary divisions, 48, 122
Pearce, A. S., 106
Pepler, G. L., 6
Permeable strata, 118
Petrology, 18
Photographic records, 124–132
  permanence of, 129, 130
Photographic Survey and Record of Surrey, *frontispiece*
Photographs, 6, 124–132

Physical characters of region, 9, 11, 13, 22, 96–104, 117, 118
  determining survey area, 22
  on transect chart, 115
Pictorial illustrations, 124–132
Picture postcards, 127
Pilgrim's Way, 125
Place names, 48, 70, 120, 123, 134
Plant ecology, 19, 104, 119
Plateau gravels, 119
Pledge, J. H., xi, 126, 129
Ponds, 19, 118
Poor Law Unions, 48, 122
Population, 20, 94, 95, 108, 109, 111, 121
Population density, 109
Population maps, 108, 109, 111
Postal service, 122
Prehistoric culture periods, 120
Prehistoric earthworks, 19, 64, 93, 120
Prehistoric man, 19, 101
Prehistory, 19, 119
Profile of river valley, 117
Publications of H.M. Ordnance Survey, 50
Public health, 20, 122
Pugh, E. C., 106

Quadrat charts, 119
Quarries, 60, 84, 101, 121, 140

Railway accessibility map, 109, 122
Railway cuttings, 101
Railway map, 108
Rainfall, 10, 12, 18, 103, 104, 115, 118
  on transect chart, 115
Rainfall Atlas, 104
Rainfall maps, 104, 118
Recreations, 20
Rectangular survey areas, 23–26
Regimen charts of rivers, 118
Regional survey colours, x, 52–57
Reigate, 141
Relief, 19, 77–92, 94, 96–98, 118
Relief maps, 96–97, 118
Relief models, 87–92, 98, 118

INDEX 149

Relief transects, 77–80, 118
Religion and religious influences, 20
Reproduction, of drawings, 46
　of Ordnance Survey maps, 36, 43, 47
River, 13, 19, 31, 98, 99, 101, 117, 118
River basins, 19, 22, 98, 117, 118
River gravels, 100, 119
Roads, 31, 108, 122
Robins, E. A., xi, 126, 129
Rochester, 31
Rock—definition, 9
Romano-British settlements, 120
Roman occupation, 19, 120
Roman roads, 18, 107, 120
Roman villas, 120
Royal Geographical Society, 24
Royal Meteorological Society, 104
Rural areas, 52, 65–76, 109
Rural districts, 29, 122
Russell, Sir E. J., 112

Salt marsh, 19
Sand dunes, 119
Saturated soils, 119
Saxon culture, 107
Saxton, Christopher, 47
Scrub, 19
Sea coasts, 19
Seasonal streams, 119
Sectarian religious organisation, 20, 123
Semi-natural vegetation, 105, 119
Shingle, 19
Shops, 20
Silhouette transects, 97, 118, 122
Smith, Ellen, 25
Social survey, 20, 122
Sociological Review, 135
Sociology, 10, 15, 108
Soil map, 118, 121
Soils, 18, 101, 106
Solid geology maps, 100, 117
South-Eastern Naturalist, xi, 6, 16
South-Eastern Union of Scientific Societies, ix, x
Sports, 20

Springs, 13, 19, 98, 115, 118, 119
Statistical tables, 93
Statistics of surface utilisation survey, 61–63
Statutory companies, 122
Stirling, 32
Stone implements, 101
Storm maps, 105, 118
Stour River, Kent, 142
Stratigraphy, 17, 18
Streams, 13, 19, 31, 98, 99, 115, 118
Subsequent streams, 99, 115
Subsoil map, 118
Sunshine, 18
Surface drainage, 98, 115
Surface utilisation survey, 48, 51–64, 76, 98, 101, 104, 110, 116, 121
Surnames, 107
Sussex, 24

Tansley, A. G., 105, 119
Tectonics, 18
Temperature, 18, 103
Tenancy of land, 121
Tertiary rocks, 99
Timber, 121
Town as centre of region, 26
Town planning, 6, 142
Traffic census, 122
Transect, 30, 77–87, 93, 103
Transect chart or diagram, 101, 112, 113, 115, 116
Transparent maps, 45, 95, 96, 101, 117, 118, 121
Transport, 20

Underground water, 19, 119
Upper Greensand, 80, 81, 114, 115
Urban areas, 18, 52, 58, 109, 121, 122

Vale of Holmesdale, 99, 115
Valley section, 116, 136
Value of land, 121
Vegetation, 10, 13, 14, 17, 19, 53, 95, 104, 119, 124, 125
　on transect chart, 115

INDEX

Vertical sections, 82, 117
Victoria County Histories, 120
Vital statistics, 122

Wandle basin, 141
Watersheds, 22, 98, 118
Water supply, 82
Water Supply Memoirs, 82, 100, 104
Water table, 119
Water undertakings, 122
Weald, 99
Weald Clay, 80, 81, 99, 114, 115
Weather, 30

Weather charts, 118
Well borings, 82, 84, 103, 117, 119
Wells, 119
Wet soils, 119
Wind, 103
Wind-roses, 118
Winsor and Newton, "Regional Survey Colours", x, 55, 96, 100
Winterbournes, 119
Woodland, 19, 95, 104, 119

Zones, of railway accessibility, 109
of the valley section, 136–142
Zoology, 19

www.ingramcontent.com/pod-product-compliance
Ingram Content Group UK Ltd.
Pitfield, Milton Keynes, MK11 3LW, UK
UKHW040657180125
453697UK00010B/226